Marine Compounds and Cancer

Special Issue Editors

Friedemann Honecker

Sergey A. Dyshlovoy

MDPI • Basel • Beijing • Wuhan • Barcelona • Belgrade

MDPI

Special Issue Editors
Friedemann Honecker
Tumor and Breast Center ZeTuP St. Gallen
Switzerland

Sergey A. Dyshlovoy
University Medical Center Hamburg-Eppendorf
Germany

Editorial Office
MDPI AG
St. Alban-Anlage 66
Basel, Switzerland

This edition is a reprint of the Special Issue published online in the open access journal *Marine Drugs* (ISSN 1660-3397) in 2017 (available at: http://www.mdpi.com/journal/marinedrugs/special_issues/marine-compounds-cancer).

For citation purposes, cite each article independently as indicated on the article page online and as indicated below:

Lastname, F.M.; Lastname, F.M. Article title. *Journal Name* **Year**, *Article number, page range.*

First Edition 2018

ISBN 978-3-03842-765-0 (Pbk)
ISBN 978-3-03842-766-7 (PDF)

Table of Contents

About the Special Issue Editors

Friedemann Honecker, MD, PhD, studied medicine in Germany and the UK between 1991 and 1999. In 1999, he started his medical and scientific career in internal medicine with a focus on Oncology and Hematology at the University Medical Center Tübingen, Germany. From 2002 to 2004, he did his PhD in experimental pathology at Rotterdam Erasmus MC in Rotterdam, NL, focusing on germ cell tumor development and mechanisms of drug resistance. From 2005 to 2013, he worked in the Department of Oncology and Hematology at Hamburg University Hospital, both as a senior consultant in Oncology and group leader of the "Laboratory of Experimental Oncology" of the University Cancer Center Hamburg (UCCH), a comprehensive cancer center. Since 2013, he has worked as an oncologist and researcher at the Tumor and Breast Center ZeTuP in St. Gallen, Switzerland. His main research interests are breast cancer and genitourinary cancers, development of new anti-cancer substances, and the treatment of elderly cancer patients. He has published over 90 original Pub-Med listed articles and reviews, and is author of numerous book chapters.

Sergey A. Dyshlovoy, PhD, trained in chemistry and biochemistry in the Russian Federation and in Germany. He started his scientific career in 2006. Between 2009 and 2012, he did his PhD in chemistry, focusing on the elucidation of structure, and mode of action of novel small-molecule marine bioactive compounds, showing anticancer activity in the in G.B. Elyakov Pacific Institute of Bioorganic Chemistry (Vladivostok, Russian Federation). Since 2012, he has been a senior researcher at both this institute and at the Far Eastern Federal University (Vladivostok). Since 2013, he has worked as a post-doctoral researcher at the University Medical Center Hamburg-Eppendorf (Hamburg, Germany) in the Department of Oncology. Since 2014, he has been a member of the Commission of Experts of the Russian Scientific Foundation. His main research interest is the investigation of mechanisms of anticancer action of novel marine compounds with a strong focus on autophagy. To date, he has published over 50 research articles.

marine drugs

MDPI

Editorial

Marine Compounds and Cancer: 2017 Updates

Sergey A. Dyshlovoy [1,2,3,*] and Friedemann Honecker [2,4,*]

1 Laboratory of Marine Natural Products Chemistry, G.B. Elyakov Pacific Institute of Bioorganic Chemistry, Far-East Branch, Russian Academy of Sciences, 690022 Vladivostok, Russia
2 Laboratory of Experimental Oncology, Department of Oncology, Hematology and Bone Marrow Transplantation with Section Pneumology, Hubertus Wald-Tumorzentrum, University Medical Center Hamburg-Eppendorf, 20246 Hamburg, Germany
3 School of Natural Sciences, Far East Federal University, 690022 Vladivostok, Russia
4 Tumor and Breast Center ZeTuP St. Gallen, CH-9006 St. Gallen, Switzerland
* Correspondence: dyshlovoy@gmail.com (S.A.D.); Friedemann.Honecker@zetup.ch (F.H.)

Received: 15 January 2018; Accepted: 22 January 2018; Published: 24 January 2018

By the end of 2017, there were seven marine-derived pharmaceutical substances that have been approved by the FDA for clinical use as drugs. Four of them are approved for the treatment of cancer, namely cytarabine (Cytosar-U®, first approved in 1969 for the treatment of leukemia), eribulin mesylate (Halaven®, first approved in 2010 for the treatment of metastatic breast cancer), brentuximab vedotin (Adcetris®, first approved in 2011 for the treatment of anaplastic large T-cell malignant lymphoma, and Hodgkin's lymphoma), and trabectidine (Yondelis®, first approved in 2015 for the treatment of soft tissue sarcoma and ovarian cancer) [1]. Additionally, a number of marine-derived substances with potent anticancer properties are currently undergoing different stages of clinical development in oncology and hematology. Among them are plinabulin, plitidepsin, glembatumumab vedotin, and lurbinectedin (all in Phase III clinical trials); depatuxizumab mafodotin, AGS-16C3F, polatuzumab vedotin, PM184, tisotumab vedotin, and enfortumab vedotin (all in Phase II clinical trials); GSK2857916, ABBV-085, ABBV-399, ABBV-221, ASG-67E, ASG-15ME, bryostatin, marizomib, and SGN-LIV1A (all in Phase I clinical trials) [1]. Additionally, there are around 1500 natural molecules that were isolated from marine organisms, for which potent in vivo biological activity has been described, and more than 10,000 different compounds that have exhibited in vitro activity [2]. Therefore, natural compounds are a rich reservoir of molecules showing promising bioactivity, which will most certainly lead to the further development of potent anticancer compounds in the future.

To document this dynamic field of research, the topical collection "Marine Compounds and Cancer" (http://www.mdpi.com/journal/marinedrugs/special_issues/marine-compounds-cancer) of the open access journal *Marine Drugs* (ISSN 1660-3397) was started in 2017, one year after the special issue under the same name had been closed [3]. In 2017, a total of nine papers in this collection—two reviews and seven research articles—were published. The collection covers both novel and previously known anticancer agents from different classes of small and high molecular compounds, exploring their chemical structures and anticancer activity. Below, a short overview of the articles published in 2017 in the topical collection "Marine Compounds and Cancer" is provided.

A comprehensive review article by Ćetković and colleagues from Croatia reviews the current knowledge of **cancer-related genes/proteins found in marine sponges**. Elucidating cancer-associated genes in such simple and ancient animals as sponges may help to understand the more complex signaling interactions in higher animals [4]. Another review article by Martínez-Poveda and colleagues from Spain describes various biological activities of **puupehenones** (named after the famous rock Puu Pehe in Hawaii)—a large family of chemical compounds initially isolated from sponges of the orders Verongida and Dictyoceratida. This review gives an update on the current knowledge and understanding of the bioactivity and biogenesis of puupehenones, and their possible therapeutic applications in human diseases, with special emphasis on cancer [5].

Among the research articles, there is a report by Sarmiento-Vizcaíno and colleagues from Spain on **paulomycin G**—a novel molecule of the paulomycin family, isolated from a deep-sea sediment-derived micromonospora matsumotoense M-412. This compound demonstrates moderate cytotoxic activity in human cancer cells, but no antibacterial or antifungal activity [6]. Sperlich and colleagues from Germany and Canada show that the marine diterpene glycosides **pseudopterosins** have the ability to block the major inflammatory signaling pathway NF-κB by inhibiting the phosphorylation of p65 and IκB in leukemia and breast cancer cells, resulting in a reduction of the pro-inflammatory cytokines IL-6, TNFα, and MCP-1. They hypothesize that pseudopterosins inhibit NF-κB through activation of the glucocorticoid receptor in triple negative breast cancer [7]. Sun and colleagues from China report anticancer activity of a **selenium-containing polysaccharide-protein complex**, isolated from the selenium-enriched algae *Ulva fasciata*. Functionally, they describe the ability of this complex to induce mitochondria-mediated apoptosis in human lung cancer cells [8]. Xin and colleagues, representing another research group from China investigating marine substances, report the results of a virtual screening of low cytotoxic and non-cytotoxic natural products for their **ability to inhibit topoisomerase I**, which they consecutively confirmed experimentally by biological assays [9]. Another interesting research project performed by Dithmer and colleagues from Germany and the UK report unexpected cytoprotective activity of **fucoidan** in uveal melanoma cells as well as pro-angiogenic properties. Thus, despite the fact that fucoidan has previously been shown to also exhibit anti-tumorigenic effects, the authors conclude that this polysaccharide has no potential as a novel therapy for this type of cancer [10]. Hegazy and colleagues from Egypt, Japan and the USA report the isolation of three new cembrene diterpenoids—**sarcoehrenbergilid A–C**—along with four known diterpenoids and one steroid. The compounds were isolated from the Red Sea soft coral *Sarcophyton ehrenbergi*. The biological studies revealed moderate cytotoxic activity of these compounds in human cancer cell lines. Interestingly, additional molecular docking studies predict a high inhibitory activity of some of the compounds against the kinase domain of the growth receptor EGFR [11]. Schirmeister and colleagues from Germany and Italy describe the isolation of a new **6-epi-plakortide H acid** along with several previously known compounds from the caribbean sponge *Plakortis halichondrioides*. The compounds exhibit potent cytotoxic activity in both sensitive and multidrug-resistant human leukemia cells [12]. Finally, Xu and colleagues from China isolated two new **brevianamides** and two new **mycophenolic acid** derivatives, along with several previously known compounds, from the deep-sea-derived fungus *Penicillium brevicompactum* DFFSCS025. Among others, one compound exhibits moderate cytotoxicity against human colon cancer HCT116 cells [13].

In summary, our updated topical collection of *Marine Drugs*, "Marine Compounds and Cancer", compiles recent results of research activities in the field of anticancer marine compounds from the year 2017. Exerting the honorable task of editing this collection, we express our gratitude to all authors who contributed. We are all looking forward to new and exciting discoveries!

Dr. Sergey A. Dyshlovoy and Dr. Friedemann Honecker,
Guest Editors of "Marine Compounds and Cancer", and Editorial Board Members of *Marine Drugs*

Conflicts of Interest: The authors declare no conflicts of interest.

References

1. Mayer, A. Marine Pharmaceutical: The Clinical Pipeline. Available online: http://marinepharmacology. midwestern.edu/clinPipeline.htm (accessed on 3 January 2018).
2. Mayer, A. The marine pharmacology and pharmaceuticals pipeline in 2017. In Proceedings of the 10th European Conference on Marine Products, Kolymbari, Crete, Greece, 3–7 September 2017.
3. Dyshlovoy, S.A.; Honecker, F. Marine compounds and cancer: Where do we stand? *Mar. Drugs* **2015**, *13*, 5657–5665. [CrossRef] [PubMed]
4. Cetkovic, H.; Halasz, M.; Herak Bosnar, M. Sponges: A reservoir of genes implicated in human cancer. *Mar. Drugs* **2018**, *16*, 20. [CrossRef] [PubMed]

5. Martinez-Poveda, B.; Quesada, A.R.; Medina, M.A. Pleiotropic role of puupehenones in biomedical research. *Mar. Drugs* **2017**, *15*, 325. [CrossRef] [PubMed]

6. Sarmiento-Vizcaino, A.; Brana, A.F.; Perez-Victoria, I.; Pérez-Victoria, I.; Martín, J.; de Pedro, N.; Cruz, M.; Díaz, C.; Vicente, F.; Acuña, J.L.; et al. Paulomycin G, a New Natural Product with Cytotoxic Activity against Tumor Cell Lines Produced by Deep-Sea Sediment Derived Micromonospora matsumotoense M-412 from the Aviles Canyon in the Cantabrian Sea. *Mar. Drugs* **2017**, *15*, 271. [CrossRef] [PubMed]

7. Sperlich, J.; Kerr, R.; Teusch, N. The marine natural product pseudopterosin blocks cytokine release of triple-negative breast cancer and monocytic leukemia cells by inhibiting NF-κB signaling. *Mar. Drugs* **2017**, *15*, 262. [CrossRef] [PubMed]

8. Sun, X.; Zhong, Y.; Luo, H.; Yang, Y. Selenium-containing polysaccharide-protein complex in se-enriched ulva fasciata induces mitochondria-mediated apoptosis in A549 human lung cancer cells. *Mar. Drugs* **2017**, *15*, 215. [CrossRef] [PubMed]

9. Xin, L.T.; Liu, L.; Shao, C.L.; Yu, R.-L.; Chen, F.-L.; Yue, S.-J.; Wang, M.; Guo, Z.-L.; Fan, Y.-C.; Guan, H.-S.; et al. Discovery of DNA Topoisomerase I Inhibitors with Low-Cytotoxicity Based on Virtual Screening from Natural Products. *Mar. Drugs* **2017**, *15*, 217. [CrossRef] [PubMed]

10. Dithmer, M.; Kirsch, A.M.; Richert, E.; Fuchs, S.; Wang, F.; Schmidt, H.; Coupland, S.E.; Roider, J.; Klettner, A. Fucoidan does not exert anti-tumorigenic effects on uveal melanoma cell lines. *Mar. Drugs* **2017**, *15*, 193. [CrossRef] [PubMed]

11. Hegazy, M.F.; Elshamy, A.I.; Mohamed, T.A.; Hamed, A.R.; Ibrahim, M.A.A.; Ohta, S.; Paré, P.W. Cembrene diterpenoids with ether linkages from sarcophyton ehrenbergi: An anti-proliferation and molecular-docking assessment. *Mar. Drugs* **2017**, *15*, 192. [CrossRef] [PubMed]

12. Schirmeister, T.; Oli, S.; Wu, H.; della Sala, G.; Costantino, V.; Seo, E.-J.; Efferth, T. Cytotoxicity of endoperoxides from the caribbean sponge plakortis halichondrioides towards sensitive and multidrug-resistant leukemia cells: Acids vs. esters activity evaluation. *Mar. Drugs* **2017**, *15*, 63. [CrossRef] [PubMed]

13. Xu, X.; Zhang, X.; Nong, X.; Wang, J.; Qi, S. Brevianamides and mycophenolic acid derivatives from the deep-sea-derived fungus penicillium brevicompactum DFFSCS025. *Mar. Drugs* **2017**, *15*, 43. [CrossRef] [PubMed]

marine drugs

Review

Pleiotropic Role of Puupehenones in Biomedical Research

Beatriz Martínez-Poveda [1], **Ana R. Quesada** [1,2] **and Miguel Ángel Medina** [1,2,*]

[1] Department of Molecular Biology and Biochemistry, Faculty of Sciences, University of Málaga, Andalucía Tech, and IBIMA; E-29071 Málaga, Spain; bmpoveda@uma.es (B.M.-P); quesada@uma.es (A.R.Q.)
[2] Unidad 741 de CIBER "de Enfermedades Raras", E-29071 Málaga, Spain
* Correspondence: medina@uma.es; Tel.: +34-9521-37132; Fax: +34-9521-32000

Received: 1 September 2017; Accepted: 16 October 2017; Published: 21 October 2017

Abstract: Marine sponges represent a vast source of metabolites with very interesting potential biomedical applications. Puupehenones are sesquiterpene quinones isolated from sponges of the orders *Verongida* and *Dictyoceratida*. This family of chemical compounds is composed of a high number of metabolites, including puupehenone, the most characteristic compound of the family. Chemical synthesis of puupehenone has been reached by different routes, and the special chemical reactivity of this molecule has allowed the synthesis of many puupehenone-derived compounds. The biological activities of puupehenones are very diverse, including antiangiogenic, antitumoral, antioxidant, antimicrobial, immunomodulatory and antiatherosclerotic effects. Despite the very important roles described for puupehenones concerning different pathologies, the exact mechanism of action of these compounds and the putative therapeutic effects in vivo remain to be elucidated. This review offers an updated and global view about the biology of puupehenones and their therapeutic possibilities in human diseases such as cancer.

Keywords: puupehenones; sponges; marine drugs; antiangiogenic; antitumoral

1. Origin and Biological Role of Puupehenones in Sponges

The need of new pharmacological approaches for the treatment of certain refractory diseases is the starting point for a growing research field in recent years. Indeed, discovery and characterization of new natural products and derivatives with potential therapeutic activity are key issues in pharmacological research. Natural products under investigation have a wide variety of origins, but some of the most important sources of candidate compounds for biomedical applications are marine organisms. Among the great biodiversity present in oceans and seas, sponges represent a real treasure for the isolation of new compounds with unique structural characteristics, due to the synthesis in these organisms of a high number of secondary metabolites. Sponges (phylum *Porifera*) are sessile and filter-feeder multicellular organisms that lack body symmetry. The soft body of the majority of sponges and the incapacity of movement make these organisms a perfect target for predators (fish, turtles and invertebrates); the adaptive strategy of sponges to such threats is the synthesis of chemical compounds that have a defensive role to deter predators [1]. The chemical nature of these compounds is very diverse, including sterols, terpenes, cyclic peptides, alkaloids, fatty acids, peroxides, amino acid derivatives (frequently halogenated) and unusual nucleosides [2].

One group of marine compounds synthesized by sponges that deserves special attention is the group of the puupehenones. Puupehenones are shikimate-derived sesquiterpene quinones whose isolation has been reported mainly from the orders *Verongida* and *Dictyoceratida*, although some compounds from this family have been identified as well in orders *Dendroceratida* and *Haplosclerida* [3]. Among all the compounds that belong to this family, puupehenone is the most representative member. It was firstly isolated and described by B. N. Ravi and colleagues in 1979, who named the compound in honor of the legendary Hawaiian princess Puupehe [4], but its absolute stereochemistry was not elucidated until 1996 [5].

Puupehenone and other related compounds exhibit very potent cytotoxic and antimicrobial activities, pointing to their possible role as defensive weapons in sponges. Apart from these detected activities that could be important in the chemical ecology of sponges, the exact role of puupehenones in sponges' biology is not fully defined, although an interesting mechanism has been proposed for puupehenone by which this metabolite could participate in the detoxification of excess of hydrogen cyanide (HCN), probably produced by sponges as chemical weapon with defensive purpose [6]. It has been reported that harvested sponges from the order *Verongida* emitted HCN when they were broken apart, and this observation correlates with the necessity of a mechanism of detoxification of this toxic compound in the sponge [7]. The easy conversion of puupehenone into its cyano-derivatives (15α-cyanopuupehenol and its oxidation product 15α-cyanopuupehenone) by the addition of hydrogen cyanide under aqueous conditions suggests a possible hydrogen–cyanide–puupehenone cycle, highlighting the putative biological function of puupehenone in the sponge's biochemical system [6].

Our group has contributed to the knowledge of puupehenones, focusing on their activity as antiangiogenic and pro-apoptotic compounds [8,9]. Their potential as antitumoral compounds makes puupehenones a very interesting family of metabolites for biomedical and pharmaceutical research. The information compiled in this review tries to provide an updated and global view about puupehenones' biology and their therapeutic possibilities.

2. Diversity and Chemical Synthesis of Puupehenones

The compounds gathered in the family of puupehenones are very diverse (Figure 1) and chemically belong to the large group of the sesquiterpene quinones. They have very characteristic structures, presenting a common tetracyclic core (a sesquiterpene unit joined to a phenolic moiety). Puupehenone, the most representative compound of this family, structurally differs from other sesquiterpene quinones because of the presence of a quinone–methide system responsible for its unique chemical behavior; it exhibits high chemical reactivity, facilitating the formation of many derived metabolites. The 1,6-Conjugated nucleophilic addition of HCN to puupehenone in the presence of water and alkaline conditions yields 15α-cyanopuupehenol and its oxidation product 15α-cyanopuupehenone [6]. Addition of oxygen nucleophiles such as acetoxy and methoxy ions to puupehenone (obtaining 15α-acetoxypuupehenol diacetate and 15α-methoxypuupehenol) has been also reported [4,10]. A large number of puupehenone-derived/related compounds, either naturally occurring or of synthetic origin, has been reported in the literature [11]. Some of them are shown in Figure 1.

Chemical synthesis of several puupehenones has been reported, using different synthesis routes and several initial compounds (Figure 2). The total synthesis of (±)-puupehenone was firstly described in 1978, when G. L. Trammel showed a method that applied acid-mediated cyclization of sesamol derivatives [12]. Twenty years later, Barrero et al. detected a lack of reproducibility in this method, and they reported the enantiospecific synthesis of (+)-puupehenone from the bicyclic diterpene (−)-sclareol, a fragrant compound extracted from clary sage flowers (*Salvia sclarea*) [13]. The same group proposed an improved method for puupehenone synthesis [14]. However, these are not the only approaches for the synthesis of puupehenone; in 2002, an alternative and shorter synthetic route starting from (+)-sclareolide was described, in which the heterocyclization needed for the synthesis of the molecule was mediated through the presence of an oxygen function at C-8 of (+)-sclareolide [15].

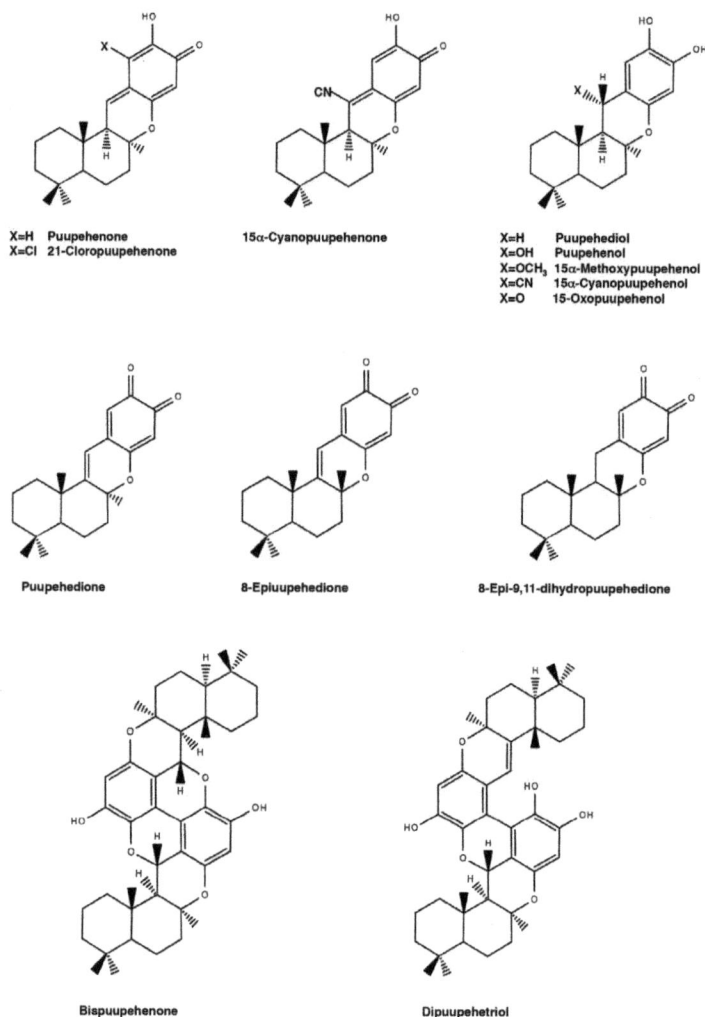

X=H Puupehenone
X=Cl 21-Cloropuupehenone

15α-Cyanopuupehenone

X=H	Puupehediol
X=OH	Puupehenol
X=OCH₃	15α-Methoxypuupehenol
X=CN	15α-Cyanopuupehenol
X=O	15-Oxopuupehenol

Puupehedione

8-Epiuupehedione

8-Epi-9,11-dihydropuupehedione

Bispuupehenone

Dipuupehetriol

Figure 1. Chemical structure of puupehenone and some derived compounds.

Different strategies targeting the synthesis of the tetracyclic core of puupehenones were further developed, providing new and improved synthetic routes. Thus, Wallace and collaborators described a three-step stereoselective reaction to access the tetracyclic core of puupehenone and 15-oxopuupehenol using methal-free radical cyclisations [16]. In addition, the construction of the tetracyclic core of puupehenone by using the Diels–Alder reaction of 2-ethenyl-1,3,3-trimethylcyclohexene with 4H-chromen-4-ones has been described [17].

In addition to the use of (−)-sclareol as a starting point for the synthesis of puupehenone [13], this easily commercially available compound has been used for the synthesis of other puupehenone-related compounds, as is the case of 15-oxopuupehenol, puupehedione, 15α-cyanopuupehenone [14], 8-epipuupehedione [18] and others [19]. Concerning the synthesis of 8-epipuupehedione, different approaches have also been described. After the report of Álvarez-Manzaneda et al. [18], the same group described the synthesis of this molecule using natural drimenol as the initial molecule [20]. Moreover, 8-epipuupehedione and puupehedione syntheses were previously described through concomitant *O*-allyl deprotection and electrocyclization of an intermediate dione molecule derived from (−)-carvone [21]. In a more recent study, Dixon and collaborators developed a scalable, divergent synthesis of meroterpenoids through the invention of a "borono–sclareolide" precursor that allowed a high-yield of (+)-chromozonarol. This intermediate was used for the subsequent syntheses of a variety of meroterpenoids, including (+)-8-epipuupehedione [22].

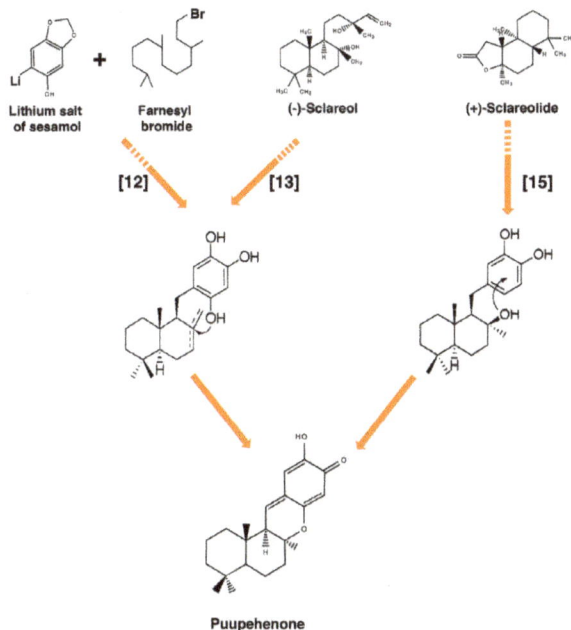

Figure 2. Chemical strategies for the synthesis of puupehenone. Scheme of the synthetic strategies reported in [12,13,15] for the synthesis of puupehenone, showing the starting compounds and the differential steps used to obtain the heterocyclic oxygen in each approach. Adapted with permission from [15].

The syntheses of (+)-chloropuupehenone, (+)-choloropuupehenol and their stereoisomers were described for the first time by Hua and collaborators. In that report, the authors investigated two synthetic routes to get (+)-chloropuupehenone, trying to improve the final yield of the molecule [23].

3. Biological Activities of Puupehenones

Puupehenones have been described as a family of compounds with very diverse and interesting biological effects. Different activities have been reported for a number of puupehenones, including antiangiogenic, antitumoral, antioxidant, antimicrobial, immunomodulatory and antiatherosclerotic effects. Here, we summarize the main information about the biological activities detected for puupehenones.

3.1. Antiangiogenic Activity

In a blind screening for the search of potential antiangiogenic compounds, puupehenone was selected due to its efficacy in inhibiting the formation of tubular-like structures on Matrigel in bovine aortic endothelial cells (BAEC) at a very low dose (3 µM) [8]. In addition to puupehenone, 11 structurally-related compounds (all of which were terpenylquinones with a labdane-type decalin ring, either natural products from marine origin or their synthetic derivatives) were evaluated, showing that some of these compounds exhibited even more potent antiangiogenic activity than puupehenone, with inhibitory doses as low as 0.37 µM for some of them. Interestingly, the 12 compounds studied showed a weak effect on BAEC cell growth, with IC_{50} values ranging from 7–45 µM, which imply that the observed effect of these compounds on the tube formation assay was not due to the inhibition of cell growth. In addition, these IC_{50} values did not differ from those obtained in tumor cell lines (human lung carcinoma, colon and pancreatic adenocarcinomas, breast carcinoma and glioblastoma cell lines), demonstrating that the effects of the assayed puupehenones on cell growth were not specific for endothelial cells. In vivo assays on chick chorioallantoic membrane (CAM) showed that puupehenone did not exhibit significant antiangiogenic effect at the assayed conditions, but in contrast, three of the related compounds studied (8-epipuupehedione, 8-epi-9,11-dihydropuupehedione and isozonarol, which is another terpenylquinone with related structure) demonstrated a very potent inhibitory effect on the CAM neovascularization, at doses of 30 nmol/CAM or even lower. Furthermore, zymographic experiments with those three antiangiogenic puupehenone-related compounds showed their activity in inhibiting the production of urokinase-type plasminogen activator (uPA) by endothelial cells. uPA is a secreted serine protease that converts plasminogen, an extracellular matrix protein, into plasmin. Related to this finding, the three selected compounds were able to inhibit the invasive capacity of endothelial cells in vitro in a modified Boyden chamber assay. Additionally, one of them (8-epipuupehedione) interfered in vitro with another important step in the angiogenic process, namely, the migration of endothelial cells. Indeed, 8-epipuupehedione, a synthetic derivative of puupehedione, was the most active compound assayed [8].

Recently, a study focused on the search for novel antiangiogenic scaffolds, pointed again to the potential of puupehenone as an inhibitor of angiogenesis [24]. In that work, 71 natural and semisynthetic compounds were filtered by a bioinformatic system attending to their novelty and druggable functionalities. Using this tool, 38 compounds were selected, tested in angiogenesis in vitro assays and screened in an angiogenesis-targeted biochemical kinase profiling. Puupehenone was one of the resulting hits, showing a high efficacy to inhibit VEGF-mediated endothelial tube-like formation in vitro. Although this compound did not progress further to the kinase profiling secondary assays, virtual screening by molecular docking of puupehenone against a panel of selected angiogenesis-related kinases suggested that glycogen synthase kinase-3 beta (GSK-3β) could be a possible kinase target of the molecule. GSK-3β is a serine/threonine kinase involved in the regulation of different cellular processes. In addition to its role in cell proliferation and inflammation, GSK-3β has been reported to play an important role in angiogenesis by inducing proangiogenic factors. The structural results relating to the binding mode of puupehenone to GSK-3β revealed that chemical modifications in the molecule could improve this binding, which offers an excellent starting point to design puupehenone-based GSK-3β inhibitors [24].

3.2. Antitumoral Effects of Puupehenones

First evidence of the antitumoral effect of puupehenone was reported by Kohmoto et al. in 1986 [25]. In that work, some values of IC_{50} for puupehenone in tumor cell lines (murine leukemia, human lung, colon and breast cancer cell lines) were shown [25], although these data were provided as ranges of values that differed in one order of magnitude (from 0.1–1 µg/mL in human lung carcinoma; from 1–10 µg/mL in human colon cancer). After these observations, more precise information about the effect of puupehenones on tumor cell lines has been reported [8,19,26,27]. Indeed, in the above-mentioned work by Castro et al., focused on the potential antiangiogenic activity of puupehenone and structurally related compounds, the authors showed the capacity of these compounds to inhibit the growth of several tumor cell lines. Reported values of IC_{50} ranged from 4 µM to more than 15 µM for the different studied compounds in a panel of cell lines [8].

As shown in Table 1, puupehenones have been tested in different tumor cell lines.

Table 1. IC$_{50}$ values of puupehenones in different cell lines.

COMPOUNDS	TESTED CELL LINES												
	A549	HT29	KB	CV1	MEL28	H116	PSN1	SKBR3	T98G	HCT8	MCF7	P388	BAEC
Puupehenone	0.4; 0.5; 0.1-1; 7	0.2; 0.5	0.5	0.5	N.D.	8	5	15 µM	>15	1-10	0.1-1	1.3; 0.25; 1	10 ± 2
Bispuupehenone	N.D.	N.D.	N.D.	N.D.	N.D.	N.D.	N.D.	N.D.	N.D.	N.D.	N.D.	N.D.	N.D.
15-oxopuupehenol	N.D.	N.D.	N.D.	N.D.	N.D.	N.D.	N.D.	N.D.	N.D.	N.D.	N.D.	N.D.	N.D.
Puupehedione	1-2	1-2	N.D.	N.D.	1	>15	>15	>15	>15	N.D.	N.D.	1	27 ± 2
Puupehediol	2.5->15	2.5	N.D.	N.D.	2.5	N.D.	N.D.	N.D.	N.D.	N.D.	N.D.	1	N.D.
Cyanopuupehenol	2	2.5; 2	N.D.	N.D.	2	N.D.	N.D.	N.D.	N.D.	N.D.	N.D.	2	N.D.
8-epipuupehedione	0.25->15	0.25	N.D.	N.D.	0.25	>15	>15	>15	>15	N.D.	N.D.	0.25	28 ± 6
8-epi-9-dihydropuupehedione	5->15	5	N.D.	N.D.	5	>15	>15	>15	>15	N.D.	N.D.	5	35 ± 7
8-epipuupehenol	1.2	1.2	N.D.	N.D.	1.2	N.D.	N.D.	N.D.	N.D.	N.D.	N.D.	1.2	N.D.
Cyanopuupehenone	5->15	1-2.5	N.D.	5	N.D.	>15	>15	>15	>15	N.D.	N.D.	5	11 ± 1
21-chloropuupehenone	0.5	0.5	N.D.	0.5	N.D.	N.D.	N.D.	N.D.	N.D.	N.D.	N.D.	0.2	N.D.
Dipuupehetriol	1	10	N.D.	0.25	N.D.	N.D.	N.D.	N.D.	N.D.	N.D.	N.D.	5	N.D.
15a-methoxypuupehenol	N.D.	N.D.	6	N.D.	N.D.	N.D.	N.D.	N.D.	N.D.	N.D.	N.D.	N.D.	N.D.
8-epipuupehediol	9 ± 1	N.D.	N.D.	N.D.	N.D.	10 ± 1	1 ± 4	>15	>15	N.D.	N.D.	N.D.	27 ± 5
8-epi-9,11-dihydropuupehediol	>15	N.D.	N.D.	N.D.	N.D.	12 ± 2	11 ± 4	10 ± 5	>15	N.D.	N.D.	N.D.	17 ± 2
Acetylpuupehenone	8 ± 4	N.D.	N.D.	N.D.	N.D.	8 ± 4	>15	10 ± 3	>15	N.D.	N.D.	N.D.	7 ± 1

IC$_{50}$ values are expressed in µg/mL, unless values in bold in the table, which correspond to µM. A549, human lung carcinoma; HT29, human colon adenocarcinoma; KB, human cervix carcinoma; CV1, monkey kidney fibroblasts; MEL28, human melanoma; H116, human colon adenocarcinoma; PSN1, human pancreatic adenocarcinoma; SKBR3, human breast carcinoma; T98G, human glioblastoma; HCT8, human colon cancer; MCF7, human breast cancer; P388, mouse leukaemia; BAEC, bovine aortic endothelium. Data compiled from [7,8,19,25-27].

However, little is known about the exact mechanism of action of these compounds to inhibit tumor cell growth. In trying to figure out its mode of action, the in vitro antitumoral activity of 8-epipuupehedione on human promyelocytic leukaemia cells (HL-60) was investigated [9]. In these cells, this compound showed a IC_{50} value lower than those obtained for other tumor and non-tumor cell lines, suggesting a certain specificity in the growth inhibition of leukaemia cells. Indeed, 8-epipuupehedione induced apoptosis in HL-60 leukaemia cells and in bovine aortic endothelial cells (BAEC), producing DNA fragmentation and effector caspase-3 activation, but these effects were not observed in the human colon adenocarcinoma cell line HCT-116. Interestingly, results in that work showed that the induction of apoptosis was stronger in the HL-60 leukaemia cell line than in BAEC. Furthermore, in leukaemia cells, 8-epipuupehedione strongly inhibited the secretion of the extracellular matrix remodeling enzyme metalloproteinase-2 (MMP2) and uPA production. This study demonstrated that 8-epipuupehedione is a potent apoptosis inductor in HL60 leukaemia cells, and a modulator of the extracellular-matrix remodeling capacity of this cell line, suggesting that in addition to its antiangiogenic activity, this compound could display a potential therapeutic effect in the treatment of promyelocytic leukaemia [9].

In a recent report, puupehenone was selected in a cell-based screen to identify natural products that were able to modulate HIF-2α in the context of renal cell carcinoma [28]. Results presented in that work showed that puupehenone inhibited HIF-2α-induced transcription of target genes. Interestingly, the data suggested that this modulatory activity might be selective for HIF-2α vs. HIF-1α. HIF-α transcription factors (HIF-1α, HIF-2α and HIF-3α) are key elements triggering the cellular response to hypoxia. Target genes of HIF encode for proteins involved in important processes, such as angiogenesis, metabolism and cell survival, that allow the cell to survive under low oxygen conditions [29]. Upregulation of the HIF pathway has been reported in several cancer types, either due to intratumoral hypoxia or to genetic mutations, and this feature correlates with a poor prognosis [30]. In renal cancer, HIF-2α has an important role in tumorigenesis [31]. Therefore, the search for compounds that inhibit this factor is an interesting antitumoral approach. This study provides a possible mechanism of action of the antitumoral effect of puupehenone in certain cancer types that rely on HIF-2α to progress, as is the case of renal cancer.

3.3. Antioxidant Activity of Puupehenones

One interesting property exhibited by some compounds of the puupehenone family is their antioxidant capacity. Puupehenone showed strong antioxidant activity in both a 2,2-diphenyl-1-picrylhydrazyl radical (DPPH) solution-based chemical assay and a 2′,7′-dichlorodihydrofluorescein diacetate (DCFH-DA) cellular-based assay, demonstrating that this compound has not only an inhibitory effect in a solution-based antioxidant assay but can also be taken up by living cells and maintain its inhibitory activity [32]. Puupehenol has been described as a potent antioxidant metabolite [33]. Isolated from a Hawaiian deep-water *Dactylospongia* sp. sponge, puupehenol and puupehenone exhibited very similar strong antioxidant activities in the ferric reducing antioxidant power (FRAP) assay [34,35].

The exact mechanism of these compounds to exert their antioxidant effect is not well-understood, but interestingly some reports have shown that puupehenone and other related compounds inhibit human lipoxygenases [36,37]. Lipoxygenases (LOX) are a family of enzymes involved in the synthesis of leukotrienes from arachidonic acid, a very important step in the inflammatory process [38]. In addition, the implication of these enzymes in the reactive oxygen species (ROS) formation has been reported [39]. In a screening focused on the search for new lipoxygenase inhibitors, puupehenone and four related compounds (chloropuupehenone, methoxypuupehenone, dimethoxypuupehenol and 20-methoxy-9-,15-ene-puupehenol) were tested as potential inhibitors of 15-LOX and 12-LOX, using an assay that directly measures the product formation of the enzymes by spectrophotometry [36]. In this study, all the five compounds exhibited an inhibitory effect against human 15-LOX, 12-LOX and 15-soybean lipoxygenase; in contrast, their inhibitory activity against 12-LOX was moderate (with IC_{50} of 8.3 µM for puupehenone). Interestingly, puupehenone was the most potent inhibitor of 15-LOX, with an IC_{50} value of 0.76 µM. The most active compound in the inhibition of 12-LOX was chloropuupehenone, with IC_{50} of 0.7 µM. In addition to 15-LOX and 12-LOX, the inhibitory effect of puupehenones against 5-LOX (a lipoxygenase isoform typically involved in inflammatory diseases such as asthma but with an emerging role in cancer [40]) has been studied, showing that puupehenone exhibited a high inhibitory activity against 5-LOX. The selectivity observed for puupehenones in these assays was diverse, but in general these compounds did not exhibit a very high selectivity against the studied lipoxygenases, with the exception of puupehenone, which presented a moderate selectivity for 5-LOX vs. 12-LOX [37].

In an assay using beef heart submitochondrial particles, the potential activity of puupehenone and five related compounds as inhibitors of the integrated electron transfer chain, in particular NADH oxidase (NOX) activity, was tested [41]. NOX enzymes are a family of proteins that transfer electrons across biological membranes. As a consequence of their activity, a superoxide ion is produced, therefore generating ROS [42]. In the work by Ciavatta et al., all the six puupehenone-related compounds assayed showed an inhibitory effect against NOX activity, with IC_{50} values that ranged from 1.3 µM (for puupehenone, the most potent inhibitor of NOX activity in this study) to 44 µM (for bispuupehenone) [41].

Altogether, these observations could partially explain the antioxidant activity of puupehenones in cells by a putative mechanism that involves lipoxygenases and NOX inhibition.

3.4. Antimicrobial Activities of Puupehenones

Since puupehenone was first isolated and described as an active compound against Gram-positive bacteria and some fungi strains [4], several studies have reported antimicrobial activity for puupehenones (including antibacterial, antifungal, antiviral and antimalarial activities). Hamann and collaborators described, in 1993, the antifungal activity of puupehenone, cyanopuupehenol, puupehedione and chloropuupehenone against *Aspergillus oryzae*, *Penicillium notatum*, *Trichophyton mentagrophytes*, *Saccharomyces cerevisiae* and *Candida albicans* [7]. In [41], puupehenone and five related compounds were tested for antifungal and antibacterial activities. In that work, puupehenone showed moderate activity against *Candida albicans* and *Staphylococcus aureus*. Similar antimicrobial activities against *S. aureus* and the fungus *Candida tropicalis* have been reported for 15α-methoxypuupehenol [27]. A potent antifungal activity for puupehenone has been reported against *Cryptococcus neoformans* and *Candida krusei* [43]. The growth of the Gram-positive bacteria *S. aureus* and *Bacillus cereus* is also inhibited by puupehenol, showing an inhibitory activity very similar to puupehenone [33].

The antituberculosis activity of puupehenones has been reported [44]. At a concentration of 12.5 µg/mL, puupehenone, 15α-cyanopuupehenol and 15-cyanopuupehenone exhibited 99%, 96% and 90% inhibition against *Mycobacterium tuberculosis*, respectively [7]. In a recent study, two puupehenone derivatives, namely, 15α-methoxypuupehenol and puupehedione, showed similar activity against *M. tuberculosis* as that previously reported for puupehenone. Interestingly, both compounds had high selectivity against dormant bacteria, which is a specific non-replicative status of the microorganism that renders a phenotype tolerant to front-line drugs during infection [45].

Puupehenones exhibit antiviral activity. Cyanopuupehenone and puupehedione showed potent antiviral activity in different infection models (more than 80% reduction in cell infection) [7]. In addition, bispuupehenone and 15-oxopuupehenol have been reported to produce moderate reduction of viral infection [46].

Interestingly, puupehenone, 15α-methoxypuupehenol and 15-oxopuupehenol have been reported to exhibit an antimalarial effect, with low IC_{50} values against different strains of *Plasmodium falciparum* [27,46].

3.5. Immunomodulatory Activity of Puupehenones

Using the mixed lymphocyte reaction (MLR) test [47], 15-oxopuupehenol, cyanopuupehenol, cyanopuupehenone, puupehenone, 21-chloropuupehenone, puupehedione and dipuupehetriol have been shown to modulate the immunological response of T cells in vitro [7,46]. In these experiments, puupehedione was the most active compound. Little is known about the immunomodulatory role of puupehenenones, since to our knowledge there has been no further research into this remarkable activity in the literature. In an interesting study focused on the use of natural products to modify covalent biomolecules that are involved in the modulation of cellular immune responses, puupehenone was attached onto [Leu^{27}]MART-1$_{26-35}$, a modified HLA-A2-associated decapeptide identified to function as an epitope for melanoma-reactive cytotoxic T lymphocytes [48]. In spite of the low affinity of the generated adduct for the HLA-A2 molecules, it was able to moderately activate interferon-γ (IFN-γ secretion in peripheral blood and tumor-infiltrating lymphocyte clones.

3.6. Puupehenones and Atherosclerosis

Recently, a potential role of puupehenones targeting atherosclerotic disease has been reported [49]. Atherosclerosis is a cardiovascular disease caused by the formation of atheroma lesions in the vessel walls of large and medium arteries. The inflammatory response is chronically activated in atherosclerosis. During the progression of the disease, atheroma lesions accumulate lipids and cholesterol transported by circulating low-density lipoprotein (LDL), but in contrast, high-density lipoprotein (HDL) can affect reverse cholesterol transport, transferring cholesterol from the lesions to the liver for its excretion [50]. In the work of Wahab et al., 19-methoxy-9,15-ene-puupehenol and 20-methoxy-9,15-ene-puupehenol have been reported to up-regulate the activity of the scavenger receptor class B Type-1 (SR-B1, a plasma membrane receptor for HDL that mediates cholesterol transfer to and from HDL) in a SR-B1 stably expressing model of a human hepatocarcinoma cell line. Due to their high efficacy, these two compounds could be considered as full agonists of the receptor, pointing to their potential effect in the reduction of atherosclerosis progression [49].

4. Final Remarks and Future Challenges

Although the number of reports found in the literature about puupehenones is not very large, the high diversity of compounds belonging to this family and the versatile and interesting biological activities reported for them (Figure 3) make puupehenones an excellent target for biomedical research.

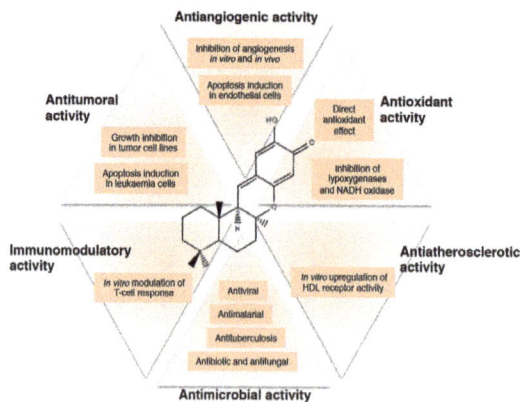

Figure 3. A summary of the multiple bio-active effects of puupehenone and derived compounds with potential therapeutic interest.

From a chemical point of view, the unique characteristics of puupehenones' structure provide an excellent scaffold for the rationale design of therapeutic agents that could improve the treatment of current resistance-associated human diseases, such as cancer. This feature has been indeed put into use for the search of new antiangiogenic and kinase inhibitor compounds [24].

Biological activities detected for puupehenones are very diverse, and their antitumoral role represents one of the most interesting effects in biomedical research. There are, however, no reports about the systemic effect of these compounds in in vivo cancer animal models, which could improve the knowledge of the antitumoral potential of puupehenones. Taking into account the antiangiogenic activity of some puupehenones [8], namely, the reported pro-apoptotic effect of 8-epipuupehedione in endothelial and leukaemia cells [9], and the inhibitory activity of puupehenone on the HIF-2α transcriptional response [28], these compounds represent a very promising putative drug against cancer. However, the exact mechanism of action of puupehenones in tumor cell lines has not been figured out yet, which opens an opportunity for future research.

Apart from their antitumoral role, the inhibitory activity of puupehenones on lipoxygeneses deserves the attention of future investigations [36,37]. In addition to ROS generation, lipoxygenases are implicated in inflammatory diseases since these enzymes catalyze the formation of eicosanoids (prostaglandins and leucotrienes) from polyunsaturated fatty acids such as linoleic and arachidonic acids [38]. Once again, the lack of in vivo data in animal models treated with puupehenones is a point to solve in subsequent studies. The same rationale could be applied to the very recent finding of the putative inhibitory role of puupehenones in atherosclerosis, a highly prevalent inflammatory disease [49]. In vivo experiments in atherosclerosis mice models, such as apolipoprotein-E knockout mice ($ApoE^{-/-}$) would shed some light on this promising therapeutic application of puupehenones.

Another important finding about puupehenones is their potential in the modulation of immune responses in T cells in vitro. This activity has been reported [7,46], but further research is needed to fully understand the molecular basis of this interesting effect.

In summary, the sponge-isolated compounds puupehenones and their synthetic derivatives represent an open field of investigation for biomedical and pharmaceutical research, and deserve the close attention of the scientific community. The lack of in vivo data about the different effects of puupehenones in several diseases could be the principal goal of future research projects on this issue, since this information could shed some light on the putative use of puupehenones as therapeutic agents.

Acknowledgments: Our experimental work is supported by grants BIO2014-56092-R (MINECO and FEDER) and P12-CTS-1507 (Andalusian Government and FEDER). The "CIBER de Enfermedades Raras" is an initiative from the ISCIII (Spain). The funders had no role in the study design, data collection and analysis, decision to publish or preparation of the manuscript.

Conflicts of Interest: The authors declare no conflict of interest.

References

1. Anjum, K.; Abbas, S.Q.; Shah, S.A.A.; Akhter, N.; Batool, S.; Hassan, S.S.U. Marine Sponges as a Drug Treasure. *Biomol. Ther. (Seoul)* **2016**, *24*, 347–362. [CrossRef] [PubMed]
2. Sipkema, D.; Franssen, M.C.R.; Osinga, R.; Tramper, J.; Wijffels, R.H. Marine sponges as pharmacy. *Mar. Biotechnol.* **2005**, *7*, 142–162. [CrossRef] [PubMed]
3. Piña, I.C.; Sanders, M.L.; Crews, P. Puupehenone Congeners from an Indo-Pacific Hyrtios Sponge. *J. Nat. Prod.* **2003**, *66*, 2–6. [CrossRef] [PubMed]
4. Ravi, B.N.; Perzanowski, H.P.; Ross, R.A.; Erdman, T.R.; Scheuer, P.J. Recent Research in Marine Natural Products: The Puupehenones. *Pure Appl. Chem.* **1979**, *51*, 1893–1900. [CrossRef]
5. Urban, S.; Capon, R.J. Absolute Stereochemistry of Puupehenone and Related Metabolites. *J. Nat. Prod.* **1996**, *59*, 900–901. [CrossRef]
6. Zjawiony, J.K.; Bartyzel, P.; Hamann, M.T. Chemistry of Puupehenone: 1,6-Conjugate Addition to Its Quinone-Methide System. *J. Nat. Prod.* **1998**, *61*, 1502–1508. [CrossRef] [PubMed]

7. Hamann, M.T.; Scheuer, P.J.; Kelly-Borges, M. Biogenetically Diverse, Bioactive Constituents of a Sponge, Order Verongida: Bromotyramines and Sesquiterpene-Shikimate Derived Metabolites. *J. Org. Chem.* **1993**, *58*, 6565–6569. [CrossRef]

8. Castro, M.E.; González-Iriarte, M.; Barrero, A.F.; Salvador-Tormo, N.; Muñoz-Chápuli, R.; Medina, M.Á.; Quesada, A.R. Study of puupehenone and related compounds as inhibitors of angiogenesis. *Int. J. Cancer* **2004**, *110*, 31–38. [CrossRef] [PubMed]

9. Martínez-Poveda, B.; Quesada, A.R.; Medina, M.Á. The anti-angiogenic 8-epipuupehedione behaves as a potential anti-leukaemic compound against HL-60 cells. *J. Cell. Mol. Med.* **2008**, *12*, 701–706. [CrossRef] [PubMed]

10. Amade, P.; Chevelot, L.; Perzanowski, H.P.; Scheuer, P.J. A Dimer of Puupehenone. *Helv. Chim. Acta* **1983**, *66*, 1672–1675. [CrossRef]

11. Gordaliza, M. Cytotoxic terpene quinones from marine sponges. *Mar. Drugs* **2010**, *8*, 2849–2870. [CrossRef] [PubMed]

12. Trammel, G.L. The total synthesis of (±)-puupehenone. *Tetrahedron Lett.* **1978**, *18*, 1525–1528. [CrossRef]

13. Barrero, A.F.; Álvarez-Manzaneda, E.J.; Chahboun, R. Enantiospecific synthesis of (+)-puupehenone from (-)-sclareol and protocatechualdehyde. *Tetrahedron Lett.* **1997**, *38*, 2325–2328. [CrossRef]

14. Alvarez-Manzaneda, E.J.; Chahboun, R.; Barranco Pérez, I.; Cabrera, E.; Alvarez, E.; Alvarez-Manzaneda, R. First enantiospecific synthesis of the antitumor marine sponge metabolite (-)-15-oxopuupehenol from (-)-sclareol. *Org. Lett.* **2005**, *7*, 1477–1480. [CrossRef] [PubMed]

15. Quideau, S.; Lebon, M.; Lamidey, A.-M. Enantiospecific Synthesis of the Antituberculosis Marine Sponge Metabolite (+)-Puupehenone. The Arenol Oxidative Activation Route. *Org. Lett.* **2002**, *4*, 3975–3978. [CrossRef] [PubMed]

16. Pritchard, R.G.; Sheldrake, H.M.; Taylor, I.Z.; Wallace, T.W. Rapid stereoselective access to the tetracyclic core of puupehenone and related sponge metabolites using metal-free radical cyclisations of cyclohexenyl-substituted 3-bromochroman-4-ones. *Tetrahedron Lett.* **2008**, *49*, 4156–4159. [CrossRef]

17. Kamble, R.M.; Ramana, M.M.V. Diels-Alder reaction of 2-ethenyl-1,3,3-trimethylcyclohexene with 4H-chromen-4-ones: A convergent approach to ABCD tetracyclic core of marine diterpenoids related to puupehenone and kampanols. *Helv. Chim. Acta* **2011**, *94*, 261–267. [CrossRef]

18. Alvarez-Manzaneda, E.J.; Chahboun, R.; Cabrera, E.; Alvarez, E.; Haidour, A.; Ramos, J.M.; Alvarez-Manzaneda, R.; Hmamouchi, M.; Bouanou, H. Diels-alder cycloaddition approach to puupehenone-related metabolites: Synthesis of the potent angiogenesis inhibitor 8-epipuupehedione. *J. Org. Chem.* **2007**, *72*, 3332–3339. [CrossRef] [PubMed]

19. Barrero, A.F.; Alvarez-Manzaneda, E.J.; Chahboun, R.; Cortés, M.; Armstrong, V. Synthesis and Antitumor Activity of Puupehedione and Related Compounds. *Tetrahedron* **1999**, *55*, 15181–15208. [CrossRef]

20. Armstrong, V.; Barrero, A.F.; Alvarez-Manzaneda, E.J.; Cortés, M.; Sepúlveda, B. An Efficient Stereoselective Synthesis of Cytotoxic 8-Epipuupehedione. *J. Nat. Prod.* **2003**, *66*, 1382–1383. [CrossRef] [PubMed]

21. Maiti, S.; Sengupta, S.; Giri, C.; Achari, B.; Banerjee, A.K. Enantiospecific synthesis of 8-epipuupehedione from (R)-(−)-carvone. *Tetrahedron Lett.* **2001**, *42*, 2389–2391. [CrossRef]

22. Dixon, D.D.; Lockner, J.W.; Zhou, Q.; Baran, P.S. Scalable, Divergent Synthesis of Meroterpenoids via "Borono-sclareolide". *J. Am. Chem. Soc.* **2012**, *134*, 8432–8435. [CrossRef] [PubMed]

23. Hua, D.H.; Huang, X.; Chen, Y.; Battina, S.K.; Tamura, M.; Noh, S.K.; Koo, S.I.; Namatame, I.; Tomoda, H.; Perchellet, E.M.; et al. Total Syntheses of (+)-Chloropuupehenone and (+)-Chloropuupehenol and Their Analogues and Evaluation of Their Bioactivities. *J. Org. Chem.* **2004**, *69*, 6065–6078. [CrossRef] [PubMed]

24. Ebrahim, H.Y.; El Sayed, K.A. Discovery of novel antiangiogenic marine natural product scaffold. *Mar. Drugs* **2016**, *14*. [CrossRef] [PubMed]

25. Kohmoto, S.; MacConnell, O.J.; Wright, A.; Koehn, F.; Thompson, W.; Lui, M.; Snader, K.M. Puupehenone, a cytotoxic metabolite from a deep water marine sponge, Stronglyophora hartmani. *J. Nat. Prod.* **1986**, *50*, 336. [CrossRef]

26. Longley, R.E.; McConnell, O.J.; Essich, E.; Harmody, D. Evaluation of Marine Sponge Metabolites for Cytotoxicity and Signal Transduction Activity. *J. Nat. Prod.* **1993**, *56*, 915–920. [CrossRef] [PubMed]

27. Bourguet-Kondracki, M.-L.; Lacombe, F.; Guyot, M. Methanol Adduct of Puupehenone, a Biologically Active Derivative from the Marine Sponge Hyrtios Species. *J. Nat. Prod.* **1999**, *62*, 1304–1305. [CrossRef] [PubMed]

28. McKee, T.C.; Rabe, D.; Bokesch, H.R.; Grkovic, T.; Whitson, E.L.; Diyabalanage, T.; Van Wyk, A.W.W.; Marcum, S.R.; Gardella, R.S.; Gustafson, K.R.; et al. Inhibition of Hypoxia Inducible Factor-2 Transcription: Isolation of Active Modulators from Marine Sponges. *J. Nat. Prod.* **2012**, *75*, 1632–1636. [CrossRef] [PubMed]

29. Semenza, G.L. Targeting HIF-1 for cancer therapy. *Nat. Rev. Cancer* **2003**, *3*, 721–732. [CrossRef] [PubMed]

30. Garber, K. New drugs target hypoxia response in tumors. *J. Natl. Cancer Inst.* **2005**, *97*, 1112–1114. [CrossRef] [PubMed]

31. Covello, K.L.; Simon, M.C.; Keith, B. Targeted replacement of hypoxia-inducible factor-1alpha by a hypoxia-inducible factor-2alpha knock-in allele promotes tumor growth. *Cancer Res.* **2005**, *65*, 2277–2286. [CrossRef] [PubMed]

32. Takamatsu, S.; Hodges, T.W.; Rajbhandari, I.; Gerwick, W.H.; Hamann, M.T.; Nagle, D.G. Marine Natural Products as Novel Antioxidant Prototypes. *J. Nat. Prod.* **2003**, *66*, 605–608. [CrossRef] [PubMed]

33. Hagiwara, K.; Garcia Hernandez, J.E.; Harper, M.K.; Carroll, A.; Motti, C.A.; Awaya, J.; Nguyen, H.Y.; Wright, A.D. Puupehenol, a potent antioxidant antimicrobial meroterpenoid from a Hawaiian deep-water Dactylospongia sp. sponge. *J. Nat. Prod.* **2015**, *78*, 325–329. [CrossRef] [PubMed]

34. Benzie, I.F.F.; Strain, J.J. The Ferric Reducing Ability of Plasma (FRAP) as a Measure of "Antioxidant Power": The FRAP Assay. *Anal. Biochem.* **1996**, *239*, 70–76. [CrossRef] [PubMed]

35. Benzie, I.F.F.; Strain, J.J. Ferric Reducing/Antioxidant Power Assay: Direct Measure of Total Antioxidant Activity of Biological Fluids and Modified Version for Simultaneous Measurement of Total Antioxidant Power and Ascorbic Acid Concentration. *Methods Enzymol.* **1999**, *299*, 15–27. [PubMed]

36. Amagata, T.; Whitman, S.; Johnson, T.A.; Stessman, C.C.; Loo, C.P.; Lobkovsky, E.; Clardy, J.; Crews, P.; Holman, T.R. Exploring Sponge-Derived Terpenoids for Their Potency and Selectivity against 12-Human, 15-Human, and 15-Soybean Lipoxygenases. *J. Nat. Prod.* **2003**, *66*, 230–235. [CrossRef] [PubMed]

37. Robinson, S.J.; Hoobler, E.K.; Riener, M.; Loveridge, S.T.; Tenney, K.; Valeriote, F.A.; Holman, T.R.; Crews, P. Using Enzyme Assays to Evaluate the Structure and Bioactivity of Sponge-Derived Meroterpenes. *J. Nat. Prod.* **2009**, *72*, 1857–1863. [CrossRef] [PubMed]

38. Mashima, R.; Okuyama, T. The role of lipoxygenases in pathophysiology; new insights and future perspectives. *Redox Biol.* **2015**, *6*, 297–310. [CrossRef] [PubMed]

39. Cho, K.J.; Seo, J.M.; Kim, J.H. Bioactive lipoxygenase metabolites stimulation of NADPH oxidases and reactive oxygen species. *Mol. Cells* **2011**, *32*, 1–5. [CrossRef] [PubMed]

40. Bishayee, K.; Khuda-Bukhsh, A. 5-Lipoxygenase Antagonist therapy: A new approach towards targeted cancer chemotherapy. *Acta Biochim.* **2013**, *45*, 709–719. [CrossRef] [PubMed]

41. Ciavatta, M.L.; Lopez Gresa, M.P.; Gavagnin, M.; Romero, V.; Melck, D.; Manzo, E.; Guo, Y.W.; van Soest, R.; Cimino, G. Studies on puupehenone-metabolites of a Dysidea sp.: Structure and biological activity. *Tetrahedron* **2007**, *63*, 1380–1384. [CrossRef]

42. Bedard, K.; Krause, K.-H. The NOX Family of ROS-Generating NADPH Oxidases: Physiology and Pathophysiology. *Physiol. Rev.* **2007**, *87*, 245–313. [CrossRef] [PubMed]

43. Xu, W.-H.; Ding, Y.; Jacob, M.R.; Agarwal, A.K.; Clark, A.M.; Ferreira, D.; Liang, Z.-S.; Li, X.-C. Puupehanol, a sesquiterpene-dihydroquinone derivative from the marine sponge Hyrtios sp. *Bioorg. Med. Chem. Lett.* **2009**, *19*, 6140–6143. [CrossRef] [PubMed]

44. El Sayed, K.A.; Bartyzel, P.; Shen, X.; Perry, T.L.; Zjawiony, J.K.; Hamann, M.T. Marine natural products as antituberculosis agents. *Tetrahedron* **2000**, *56*, 949–953. [CrossRef]

45. Felix, C.R.; Gupta, R.; Geden, S.; Roberts, J.; Winder, P.; Pomponi, S.A.; Diaz, M.C.; Reed, J.K.; Wright, A.E.; Rohde, K.H. Selective killing of dormant Mycobacterium tuberculosis by marine natural products. *Antimicrob. Agents Chemother.* **2017**, *61*. [CrossRef]

46. Nasu, S.S.; Yeung, B.K.S.; Hamann, M.T.; Scheuer, P.J. Puupehenone-Related Metabolites from Two Hawaiian Sponges, Hyrtios spp. *J. Org. Chem.* **1995**, *60*, 7290–7292. [CrossRef]

47. Meo, T. The MLR test in the mouse. In *Immunological Methods*; Lefkovits, I., Pernis, B., Eds.; Academic Press: New York, NY, USA, 1979; pp. 227–239, ISBN 0-12-442750-2.

48. Douat-Casassus, C.; Marchand-Geneste, N.; Diez, E.; Aznar, C.; Picard, P.; Geoffre, S.; Huet, A.; Bourguet-Kondracki, M.-L.; Gervois, N.; Jotereau, F.; et al. Covalent modification of a melanoma-derived antigenic peptide with a natural quinone methide. Preliminary chemical, molecular modelling and immunological evaluation studies. *Mol. Biosyst.* **2006**, *2*, 240–249. [CrossRef] [PubMed]

49. Wahab, H.A.; Pham, N.B.; Muhammad, T.S.T.; Hooper, J.N.A.; Quinn, R.J. Merosesquiterpene congeners from the Australian Sponge Hyrtios digitatus as potential drug leads for atherosclerosis disease. *Mar. Drugs* **2017**, *15*, 6. [CrossRef] [PubMed]

50. Libby, P.; Ridker, P.M.; Hansson, G.K. Progress and challenges in translating the biology of atherosclerosis. *Nature* **2011**, *473*, 317–325. [CrossRef] [PubMed]

marine drugs

MDPI

Article

Paulomycin G, a New Natural Product with Cytotoxic Activity against Tumor Cell Lines Produced by Deep-Sea Sediment Derived *Micromonospora matsumotoense* M-412 from the Avilés Canyon in the Cantabrian Sea

Aida Sarmiento-Vizcaíno [1], Alfredo F. Braña [1], Ignacio Pérez-Victoria [2], Jesús Martín [2], Nuria de Pedro [2], Mercedes de la Cruz [2], Caridad Díaz [2], Francisca Vicente [2], José L. Acuña [3], Fernando Reyes [2,*], Luis A. García [4] and Gloria Blanco [1,*]

[1] Departamento de Biología Funcional, Área de Microbiología, and Instituto Universitario de Oncología del Principado de Asturias, Universidad de Oviedo, 33006 Oviedo, Spain; UO209983@uniovi.es (A.S.-V.); afb@uniovi.es (A.F.B.)
[2] Fundación MEDINA, Centro de Excelencia en Investigación de Medicamentos Innovadores en Andalucía, Avda, del Conocimiento 34, Parque Tecnológico de Ciencias de la Salud, E-18016 Granada, Spain; ignacio.perez-victoria@medinaandalucia.es (I.P.-V.); jesus.martin@medinaandalucia.es (J.M.); ndepedro@lifelength.com (N.d.P.); mercedes.delacruz@medinaandalucia.es (M.d.l.C.); caridad.diaz@medinaandalucia.es (C.D.); francisca.vicente@medinaandalucia.es (F.V.)
[3] Departamento de Biología de Organismos y Sistemas, Área de Ecología, Universidad de Oviedo, 33006 Oviedo, Spain; acuna@uniovi.es
[4] Departamento de Ingeniería Química y Tecnología del Medio Ambiente, Área de Ingeniería Química, Universidad de Oviedo, 33006 Oviedo, Spain; luisag@uniovi.es
* Correspondences: fernando.reyes@medinaandalucia.es (F.R.); gbb@uniovi.es (G.B.); Tel.: +34-958-993-965 (F.R.); +34-985-103-205 (G.B.)

Received: 15 June 2017; Accepted: 23 August 2017; Published: 28 August 2017

Abstract: The present article describes a structurally novel natural product of the paulomycin family, designated as paulomycin G (**1**), obtained from the marine strain *Micromonospora matsumotoense* M-412, isolated from Cantabrian Sea sediments collected at 2000 m depth during an oceanographic expedition to the submarine Avilés Canyon. Paulomycin G is structurally unique since—to our knowledge—it is the first member of the paulomycin family of antibiotics lacking the paulomycose moiety. It is also the smallest bioactive paulomycin reported. Its structure was determined using HRMS and 1D and 2D NMR spectroscopy. This novel natural product displays strong cytotoxic activities against different human tumour cell lines, such as pancreatic adenocarcinoma (MiaPaca_2), breast adenocarcinoma (MCF-7), and hepatocellular carcinoma (HepG2). The compound did not show any significant bioactivity when tested against a panel of bacterial and fungal pathogens.

Keywords: paulomycins; *Micromonospora*; antitumor; Cantabrian Sea-derived actinobacteria

1. Introduction

Paulomycins are glycosylated natural products featuring a pauloate residue with pharmacological interest due to their antibiotic activities. Paulomycins A and B are antibiotics with very potent activity against Gram-positive bacteria (*Staphylococcus aureus* and *Bacillus cereus*) and are of therapeutic use in the treatment of gonococcal and *Chlamydia* infections [1]. Initially described in *Streptomyces paulus* [2] and later in *Streptomyces albus* J1074 [3], a series of paulomycins with various modifications at the two-carbon branched chain of paulomycose were subsequently isolated from these *Streptomyces* species [3–5]. The biosynthetic pathway of paulomycins is also the subject of active research [6,7], and no chemical synthesis has been reported.

In oceans, sediments are one of the most-studied marine sources for actinobacterial isolation [8,9]. Previous work in the Cantabrian Sea (Biscay Bay, Northeast Atlantic), have revealed that bioactive actinobacteria—mainly *Streptomyces* species—are associated to corals and other invertebrates living up to 4700 m depth in the submarine Avilés Canyon [10–12]. Actinobacteria displaying a wide repertoire of chemically diverse secondary metabolites with different antibiotic or antitumor activities have been isolated from coral reefs ecosystems from the Avilés Canyon [13]. Recently, new natural products with antibiotic and cytotoxic activities have been reported in this Canyon [14,15]. Paulomycins A and B have been reported to be produced by a ubiquitous *Streptomyces albidoflavus* strain widely distributed among terrestrial, marine, and atmospheric environments in the Cantabrian Cornice [12].

Herein, we report the discovery of a novel natural product, paulomycin G (**1**), obtained from *Micromonospora matsumotoense* M-412, isolated from deep sea sediments collected at 2000 m depth during an oceanographic expedition to the submarine Avilés Canyon. The presence of a new paulomycin not previously reported was identified in the extract by LC-UV-MS and LC-HRMS chemical dereplication [16], and further efforts were focused on the isolation, structural elucidation, and biological properties of this new molecule. Paulomycin G is also the first member of the family displaying strong cytotoxic activity against different human tumour cell lines, such as pancreatic adenocarcinoma (MiaPaca_2), breast adenocarcinoma (MCF-7), and hepatocellular carcinoma (HepG2).

2. Results and Discussion

2.1. Taxonomy and Phylogenetic Analysis of the Strain

The 16S rDNA of producing strain M-412 was amplified by polymerase chain reaction (PCR) and sequenced. After Basic Logic Alignment Search Tool (BLAST) sequence comparison, strain M-412 showed 100% identity to *Micromonospora matsumotoense* (Accession number NR_025015); thus, this strain was designated as *Micromonospora matsumotoense* M-412 (EMBL Sequence number LT627194). The phylogenetic tree generated by a neighbour-joining method based on 16S rDNA sequence clearly revealed the evolutionary relationship of strain M-412 with a group of known *Micromonospora* species (Figure 1). To our knowledge, all known paulomycin compounds have only been produced by *Streptomyces*; thus, paulomycin G is the first member of the family produced by a *Micromonospora* species.

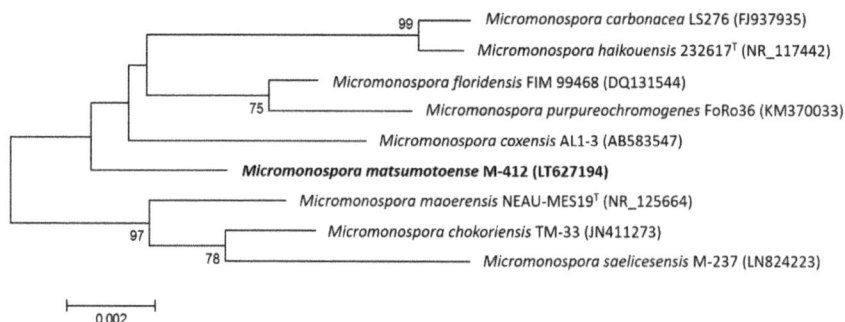

Figure 1. Neighbour-joining phylogenetic tree obtained by distance matrix analysis of 16S rDNA sequences, showing *Micromonospora matsumotoense* M-412 position and most closely related phylogenetic neighbours. Numbers on branch nodes are bootstrap values (1000 resamplings; only values >70% are given). Bar indicates 0.2% sequence divergence.

2.2. Structure Determination

Compound **1** was isolated as a pale yellow solid. LC-UV-HRMS analysis of the isolated sample revealed a purity of 83.5% (UV at 210 nm), and indicated the presence of a major impurity in the sample (16.5%), which was identified as the dehydrated paulomycinone **2** based on its UV and HRMS spectra (see Figures S1, S4 and S5). Dehydration of paulomycins has been described to occur easily, even when leaving the compounds in solution in aqueous media at neutral pH, and is therefore difficult to avoid [17]. The major compound of the mixture, paulomycin G, had a molecular formula of $C_{20}H_{22}N_2O_{11}S$ according to ESI-TOF MS measurements (m/z 499.1015 [M + H]$^+$, calcd. for $C_{20}H_{23}N_2O_{11}S^+$, 499.1017) and the presence of 20 signals in its ^{13}C NMR spectrum. Its UV spectrum displayed maxima at 238, 276, and 320 nm, in agreement with a paulomycin-like structure [2,3]. An intense absorption band at 2041 cm^{-1} in its IR spectrum confirmed the presence of an isothiocyanate group in the molecule, present in the pauloate moiety of all paulomycins. NMR spectra (Table 1) confirmed the presence of this pauloate moiety, with signals for a methyl group (δ_H 1.89, δ_C 14.6 ppm) coupled to an sp^2 proton (δ_H 6.71, δ_C 136.9 ppm) in the COSY spectrum. HMBC correlations from the latter proton to carbons at δ_C 159.8 (α,β-unsaturated carbonyl group C1''), 122.3 ppm (sp^2 quaternary carbon C2''), 141.6 ppm (isothiocyanate carbon C5'', four-bond distance correlation), and 14.6 ppm (methyl C4'') completed the structural assignment of the pauloate moiety. Additionally, signals for an aliphatic methylene at δ_H 3.22 and 3.17 that correlated in the HMBC spectrum with carbon signals at δ_C 159.4 (C3), 188.8 (C4), 77.5 (C6 and C8), and 197.5 (C7), and the presence of two additional signals at δ_C 169.0 and 99.1 ppm accounted for the presence in the molecule of the 6-substituted 2-amino-5-hydroxy-3,6-dioxocyclohex-1-enecarboxylic acid substructure present in the paulomycin family of compounds. On the other hand, signals for five oxygenated methines at δ_H 3.69, 3.61, 5.29, 4.51, and 3.79 ppm, and one aliphatic methyl group at 0.88 ppm conformed a spin system according to COSY correlations that accounted for ring B in the structure of the molecule. Connection between carbon C6 of ring A and carbon C8 in ring B was additionally confirmed via HMBC correlations observed between H8 and C5, C6, and C7. Finally, signals for an acetyl functionality were observed (δ_H 2.10 ppm, δ_C 170.1, and 20.8 ppm). This acetyl group was placed at C10 based on HMBC correlations of H2' and H10 to the carbonyl carbon C1' and the low field chemical shift of H10. The pauloate moiety was similarly placed at C11 based on an HMBC correlation observed between H11 and C1''. The relative configuration proposed around the chiral centres in ring B was based on the large coupling constants observed between H8 and H9 (9.8 Hz) and between H11 and H12 (9.9 Hz), indicating the axial orientation of all these protons. The two small coupling constants measured for H10 strongly suggested an equatorial orientation for this proton, and finally, correlations observed in the ROESY spectrum between H10, H9, and H11, and between H8 and H12, proved a relative configuration for ring B as depicted in Figure 2. Based on biogenetic considerations, the absolute configuration was assumed to be the same as in other compounds of the paulomycin series.

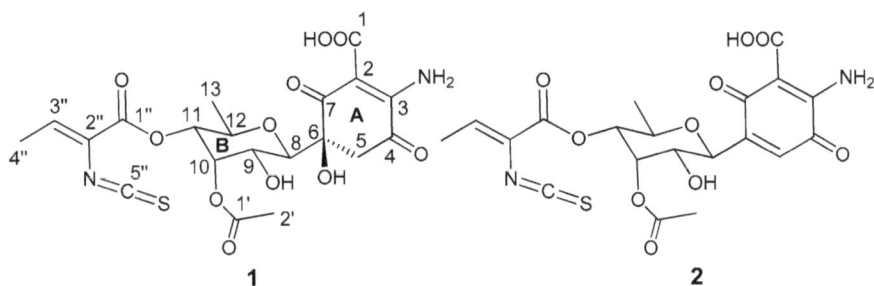

Figure 2. Chemical structures of paulomycin G (**1**) and compound **2**.

Paulomycin G is a novel natural product, structurally unique since it constitutes the first member of the paulomycin family of antibiotics lacking the paulomycose moiety and having a methyl group at position C-13. It is also the smallest bioactive paulomycin reported to-date.

Table 1. ^{1}H and ^{13}C NMR (500 and 125 MHz in DMSO-d_6) data for compound 1.

Position	δ ^{13}C	δ (^{1}H), (Mult, J in Hz)
1	169.0	-
2	99.1	-
3	159.4	-
4	188.8	-
5	47.7	3.22 (d, 16.0), 3.17 (d, 16.0)
6	77.5	-
7	197.5	-
8	77.5	3.69 (d, 9.9)
9	67.0	3.61 (br dt, 9.7, 3.3)
10	69.8	5.29 (dd, 2.6, 2.6)
11	73.4	4.51 (dd, 9.9, 2.6)
12	69.9	3.79 (dq, 9.9, 6.2)
13	16.3	0.88 (d, 6.2)
1'	170.1	-
2'	20.8	2.10 (s)
1''	159.8	-
2''	122.3	-
3''	136.9	6.71 (quart., 7.1)
4''	14.6	1.89 (d, 7.1)
5''	141.6	-
NH$_2$ (3)	-	9.71 (br s), 9.35 (br s)
OH (1)	-	14.21 (br s)
OH (6)	-	5.43 (s)
OH (9)	-	5.78 (d, 4.4)

2.3. Cytotoxic Activity of Paulomycin G

Cytotoxic activity was observed for the compound against human breast adenocarcinoma (MCF-7), pancreatic adenocarcinoma (MiaPaca_2), and hepatocellular carcinoma (HepG2) cell lines (Table 2). Considering the lack of biological activity reported for compounds having the quinone moiety present in compound **2** [18], it is reasonable to assume that the biological activity reported herein for the isolated sample is mostly due to paulomycin G. Paulomycin B did not display any cytotoxic activity against the three cell lines when tested in parallel. Figure 3 represents the dose-response curves of paulomycins B and G against the different tumour cell lines.

Table 2. Cytotoxic activity of paulomycins B and G against different tumour cell lines.

Cell line	Paulomycin G (IC$_{50}$ µM)	Paulomycin B (IC$_{50}$ µM)
HepG2	4.30 ± 0.42	>36
MCF-7	1.58 ± 0.12	>36
MiaPaca_2	2.70 ± 0.25	>36

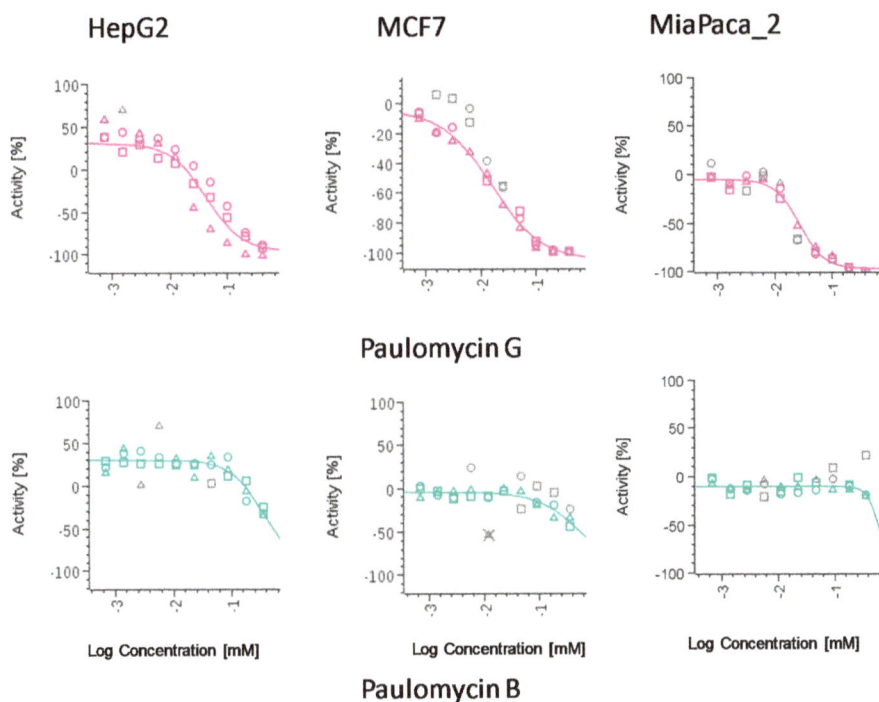

Figure 3. Dose-response curves of compounds against human breast adenocarcinoma (MCF-7), pancreatic adenocarcinoma (MiaPaca_2), and hepatocellular carcinoma (HepG2) cell lines. Compounds were tested per triplicate and the obtained results are indicated with triangles, rounds, and squares in every picture.

2.4. Antimicrobial Activity of Paulomycin G

The antimicrobial activity of Paulomycin G was tested against a panel of pathogenic bacteria and fungi, including Gram-negative (*Pseudomonas aeruginosa*, *Acinetobacter baumannii*, *Escherichia coli*, and *Klebsiella pneumoniae*) and Gram-positive bacteria (methicillin-resistant *Staphylococcus aureus*, MRSA) and fungi (*Aspergillus fumigatus* and *Candida albicans*). Paulomycin B was tested in parallel against the same panel of pathogens. One of the *E. coli* strains tested (MB5746) was the only pathogen whose growth was inhibited by the action of paulomycin G, with an MIC_{90} of 38 µg/mL. Paulomycin B displayed activity against *E. coli* MB5746 and MRSA, with MIC_{90} values of 4.5 and 50 µg/mL, respectively.

3. Materials and Methods

3.1. General Experimental Procedures

Semipreparative HPLC was performed with an Alliance chromatographic system (Waters Corporation, Mildford, MA, USA) and an Atlantis C18 column (10 µm, 10 × 150 mm, Waters). For UPLC analysis an Acquity UPLC equipment (Waters) with a BEH C18 column (1.7 µm, 2.1 × 100 mm, Waters) was used. Optical rotations were determined with a JASCO P-2000 polarimeter (JASCO Corporation, Tokyo, Japan). IR spectrum was measured with a JASCO Fourier transform infrared (FT/IR)-4100 spectrometer (JASCO Corporation) equipped with a PIKE MIRacle™ single reflection ATR accessory. NMR spectra were recorded on a Bruker Avance III spectrometer (500 and 125 MHz for ^{1}H and ^{13}C NMR, respectively) equipped with a 1.7 mm TCI MicroCryoProbe^{TM} (Bruker Biospin, Fällanden, Switzerland), using the signal of the residual solvent as internal reference (δ_H 2.50 and δ_C 39.5 ppm for DMSO-d_6). ESI-TOF MS spectra were acquired with a Bruker maXis QTOF spectrometer (Bruker Daltonik GmbH, Bremen, Germany).

3.2. Microorganism and Fermentation Conditions

Strain M-412 was isolated from a deep-sea sediment sample collected from the Cantabrian Sea at a depth of 2000 m, as previously described [13]. GHSA medium (1% glucose, 1% soy bean flour, 0.05% yeast extract, 2.1% MOPS, 0.06% $MgSO_4 \cdot 7H_2O$, 0.2% of a trace elements solution from R5A medium [19], pH 6.8) was selected for paulomycin G production. After autoclaving, the medium was supplemented with 0.4% of a 5 M solution $CaCl_2 \cdot 2H_2O$ and 3% DMSO. 20 Erlenmeyer flasks (250 mL), each containing 50 mL of GHSA medium, were inoculated with spores and incubated in an orbital shaker at 28 °C and 250 rpm during 7 days.

3.3. Isolation and Purification of Paulomycin G

The culture broths were centrifuged, and the pellets were extracted with ethyl acetate acidified with 1% formic acid. The supernatants were filtered and applied to a solid-phase extraction cartridge (Sep-Pak Vac C18, 10 g, Waters) that was eluted using a gradient of methanol and 0.05% TFA in water from 0 to 100% methanol in 60 min, at a flow rate of 5 mL/min. Fractions were collected every 5 min and analysed by UPLC using chromatographic conditions previously described [5]. A peak corresponding to an unknown paulomycin was detected in fractions eluting between 40 and 50 min. These fractions were pooled, partially dried in vacuo, and applied to a solid-phase extraction cartridge (Sep-Pak Vac C18, 2 g, Waters). The cartridge was washed with water and the retained compounds were eluted with methanol and dried in vacuo. The residue was subsequently redissolved in a small volume of acetonitrile and DMSO (2:1). The same peak of unknown paulomycin was also found in the organic extract of the culture pellets, which was dried and redissolved as above. Paulomycin G was eventually purified by semipreparative HPLC using an Atlantis C18 column (10 μm, 10 × 150 mm, Waters) in two isocratic elution steps, employing a mixture of 50% acetonitrile and water in the first step and 55% methanol and water in the second step, with a flow of 5 mL/min. In both cases, the solution containing the collected peak was evaporated and finally lyophilized, resulting in 2.7 mg of compound **1** (83.5% purity according to LC-UV analysis at 210 nm).

Paulomycin G (**1**). pale yellow solid; $[\alpha]_D^{20}$ +10.6° (*c* 0.18, MeOH); UV (DAD) λ_{max} 238, 276 y 320 nm; IR (ATR) ν_{max} 3359, 3229, 2979, 2933, 2041, 1736, 1695, 1626, 1571, 1442, 1381, 1260, 1227, 1129, 1025, 909, 751 cm^{-1}; for ^{1}H and ^{13}C NMR data see Table 1; ESI-TOF MS m/z 521.0828 [M + Na]$^+$ (calcd. for $C_{20}H_{22}N_2O_{11}SNa^+$, 521.0837) 516.1278 [M + NH$_4$]$^+$ (calcd. for $C_{20}H_{26}N_3O_{11}S^+$, 516.1283), 499.1015 [M + H]$^+$ (calcd. $C_{20}H_{23}N_2O_{11}S^+$, 499.1017).

3.4. Phylogenetic Analysis (Taxonomy) of the Producer Microorganism

Phylogenetic analysis based on 16S rRNA sequences was performed with strain *Micromonospora matsumotoense* M-412 using MEGA version 6.0 [20] after multiple alignment of data by Clustal Omega [21]. Distances (distance options according to the Kimura two-parameter model [22]) and clustering with the neighbour-joining method [23] were evaluated using bootstrap values based on 1000 replications [24].

3.5. Cytotoxic Activity of Compound **1**

The MTT (3-(4,5-dimethylthiazol-2-yl)-2,5-diphenyltetrazolium bromide) colorimetric assay—which measures mitochondrial metabolic activity—was performed using three tumour cell lines obtained from the ATCC, namely human breast adenocarcinoma (MCF-7), pancreatic adenocarcinoma (MiaPaca_2), and hepatocellular carcinoma (HepG2), using previously-described methodology [15]. Methyl methanesulphonate at a concentration of 8 mM and 0.5% DMSO in water were used as positive and negative controls, respectively.

3.6. Antimicrobial Activity of Compound **1**

Antibacterial and antifungal activity tests of paulomycins G and B were performed against the pathogenic strains *P. aeruginosa* PAO1, *Acinetobacter baumannii* MB5973, *Escherichia coli* MB5746 and MB2884, *K. pneumoniae* ATCC700603, methicillin-resistant *Staphylococcus aureus* MB5393, *Aspergillus fumigatus* ATCC46645, and *Candida albicans* ATCC64124, as previously described [25].

4. Conclusions

In summary, a new member of the paulomycin family that we have designated as paulomycin G has been obtained from cultures of *Micromonospora matsumotoense* M-412, isolated from deep-sea sediments collected at 2000 m depth in the submarine Avilés Canyon. Paulomycin G is a novel natural product, structurally unique since—to our knowledge—it is the first member of the paulomycin family of antibiotics lacking the paulomycose moiety, being the smallest bioactive paulomycin reported. This new natural product displayed strong cytotoxic activity against human cancer cell lines such as pancreatic adenocarcinoma (MiaPaca_2), breast adenocarcinoma (MCF-7), and hepatocellular carcinoma (HepG2). Based on its cytotoxic activities, paulomycin G deserves to be considered as a candidate to perform further studies assessing its anticancer potential. Besides its unique structural features, paulomycin G might also be of interest in biosynthetic studies and useful for future paulomycin biosynthesis research or as a core structure in the generation of novel derivatives through combinatorial biosynthesis to generate structural diversity in the paulomycin family. Additionally, the isolation of a compound of the paulomycin family from a non-*Streptomyces* actinomycete is particularly noteworthy and suggests horizontal gene transfer between *Micromonospora* and *Streptomyces* species.

Supplementary Materials: The following are available online at www.mdpi.com/1660-3397/15/9/271/s1. Figure S1: HPLC trace of the sample isolated; Figure S2: UV spectrum of compound 1; Figure S3: ESI-TOF spectra of compound 1; Figure S4: UV spectrum of compound 2; Figure S5: ESI-TOF spectra of compound 2; Figure S6: ^{1}H NMR spectrum (DMSO-*d6*, 500 MHz) of compound 1; Figure S7: ^{13}C NMR spectrum (DMSO-*d6*, 125 MHz) of compound 1; Figure S8: COSY spectrum (DMSO-*d6*) of compound 1; Figure S9: HSQC spectrum (DMSO-*d6*) of compound 1; Figure S10: HMBC spectrum (DMSO-*d6*) of compound 1; Figure S11: ROESY spectrum (DMSO-*d6*) of compound 1; Figure S12: Picture of Micromonospora matsumotoense M-412.

Acknowledgments: This study was financially supported by Gobierno del Principado de Asturias (SV-PA-13-ECOEMP-62). The polarimeter, IR, and NMR equipment used in this work were acquired via grants for scientific and technological infrastructures from the Ministerio de Ciencia e Innovación (Grants Nos. PCT-010000-2010-4 (NMR), INP-2011-0016-PCT-010000 ACT6 (polarimeter and IR)). The authors are grateful to all participants in the BIOCANT3 oceanographic expedition (DOSMARES) where *Micromonospora matsumotoense* M-412 was isolated. We are also grateful to J.A. Guijarro and J.L. Caso for continuous support. This is a contribution of the Asturias Marine Observatory.

Author Contributions: Aida Sarmiento-Vizcaíno isolated the strain and performed the phylogenetic analysis. Alfredo F. Braña and Aida Sarmiento-Vizcaíno identified and performed the purification of the compound. Ignacio Pérez-Victoria, Jesús Martín, and Fernando Reyes performed the structural elucidation of the compound. Nuria de Pedro, Mercedes de la Cruz, Caridad Díaz and Francisca Vicente performed the biological tests. José L. Acuña led the oceanographic expedition in which sediment samples were collected. Gloria Blanco and Fernando Reyes wrote the paper which was revised and approved by all the authors. Gloria Blanco and Luis A. García led and coordinated the research.

Conflicts of Interest: The authors declare no conflict of interest.

References

1. Novak, E. Treating *Chlamydia* Infections with Paulomycin. Patent PCT/US1987/002420, 21 April 1988.
2. Argoudelis, A.D.; Brinkley, T.A.; Brodasky, T.F.; Buege, J.A.; Meyer, H.F.; Mizsak, S.A. Paulomycins A and B. Isolation and characterization. *J. Antibiot.* **1982**, *35*, 285–294. [CrossRef] [PubMed]
3. Majer, J.; Chater, K. *Streptomyces albus* G produces an antibiotic complex identical to paulomycins A and B. *J. Gen. Microbiol.* **1987**, *133*, 2503–2507. [CrossRef] [PubMed]

4. Argoudelis, A.D.; Baczynskyj, L.; Haak, W.J.; Knoll, W.M.; Mizsak, S.A.; Shilliday, F.B. New paulomycins produced by *Streptomyces paulus*. *J. Antibiot.* **1988**, *41*, 157–169. [CrossRef] [PubMed]

5. Braña, A.F.; Rodríguez, M.; Pahari, P.; Rohr, J.; García, L.A.; Blanco, G. Activation and silencing of secondary metabolites in *Streptomyces albus* and *Streptomyces lividans* after transformation with cosmids containing the thienamycin gene cluster from *Streptomyces cattleya*. *Arch. Microbiol.* **2014**, *196*, 345–355. [CrossRef] [PubMed]

6. Li, J.; Xie, Z.; Wang, M.; Ai, G.; Chen, Y. Identification and analysis of the paulomycin biosynthetic gene cluster and titer improvement of the paulomycins in *Streptomyces paulus* NRRL 8115. *PLoS ONE* **2015**, *10*, e0120542. [CrossRef] [PubMed]

7. González, A.; Rodríguez, M.; Braña, A.F.; Méndez, C.; Salas, J.A.; Olano, C. New insights into paulomycin biosynthesis pathway in *Streptomyces albus* J1074 and generation of novel derivatives by combinatorial biosynthesis. *Microb. Cell Fact.* **2016**, *15*. [CrossRef] [PubMed]

8. Ward, A.C.; Bora, N. Diversity and biogeography of marine actinobacteria. *Curr. Opin. Microbiol.* **2006**, *9*, 279–286. [CrossRef] [PubMed]

9. Hameş-Kocabaş, E.E.; Uzel, A. Isolation strategies of marine-derived actinomycetes from sponge and sediment samples. *J. Microbiol. Methods* **2012**, *88*, 342–347. [CrossRef]

10. Braña, A.F.; Fiedler, H.P.; Nava, H.; González, V.; Sarmiento-Vizcaíno, A.; Molina, A.; Acuña, J.L.; García, L.A.; Blanco, G. Two *Streptomyces* Species Producing Antibiotic, Antitumor, and Anti Inflammatory Compounds Are Widespread Among Intertidal Macroalgae and Deep Sea Coral Reef Invertebrates from the Central Cantabrian Sea. *Microb. Ecol.* **2015**, *69*, 512–524. [CrossRef] [PubMed]

11. Sarmiento-Vizcaíno, A.; González, V.; Braña, A.F.; Molina, A.; Acuña, J.L.; García, L.A.; Blanco, G. *Myceligenerans cantabricum* sp. *nov.*, a barotolerant actinobacterium isolated from a deep cold water coral. *Int. J. Syst. Evol. Microbiol.* **2015**, *65*, 1328–1334. [PubMed]

12. Sarmiento-Vizcaíno, A.; Braña, A.F.; González, V.; Nava, H.; Molina, A.; Llera, E.; Fiedler, H.P.; Rico, J.M.; García-Flórez, L.; Acuña, J.L.; et al. Atmospheric Dispersal of Bioactive *Streptomyces albidoflavus* Strains Among Terrestrial and Marine Environments. *Microb. Ecol.* **2016**, *71*, 375–386. [CrossRef] [PubMed]

13. Sarmiento-Vizcaíno, A.; González, V.; Braña, A.F.; Palacios, J.J.; Otero, L.; Fernández, J.; Molina, A.; Kulik, A.; Vázquez, F.; Acuña, J.L.; et al. Pharmacological potential of phylogenetically diverse Actinobacteria isolated from deep-sea coral ecosystems of the submarine Avilés Canyon in the Cantabrian Sea. *Microb. Ecol.* **2017**, *73*, 338–352. [CrossRef] [PubMed]

14. Braña, A.F.; Sarmiento-Vizcaíno, A.; Pérez-Victoria, I.; Otero, L.; Fernández, J.; Palacios, J.J.; Martín, J.; de la Cruz, M.; Díaz, C.; Vicente, F.; et al. Branimycins B and C, Antibiotics Produced by the Abyssal Actinobacterium *Pseudonocardia carboxydivorans* M-227. *J. Nat. Prod.* **2017**, *80*, 569–573. [CrossRef] [PubMed]

15. Braña, A.F.; Sarmiento-Vizcaíno, A.; Osset, M.; Pérez-Victoria, I.; Martín, J.; de Pedro, N.; de la Cruz, M.; Díaz, C.; Vicente, F.; Reyes, F.; et al. Lobophorin K, a New Natural Product with Cytotoxic Activity Produced by *Streptomyces* sp. M-207 Associated with the Deep-Sea Coral *Lophelia pertusa*. *Mar. Drugs* **2017**, *15*, 144. [CrossRef] [PubMed]

16. Pérez-Victoria, I.; Martín, J.; Reyes, F. Combined LC/UV/MS and NMR strategies for the dereplication of marine natural products. *Planta Med.* **2016**, *82*, 857–871. [CrossRef] [PubMed]

17. Wiley, P.F.; Mizsak, S.A.; Baczynskyj, L.; Argoudelis, A.D.; Duchamp, D.J.; Watt, W. The Structure and Chemistry of Paulomycin. *J. Org. Chem.* **1986**, *51*, 2493–2499. [CrossRef]

18. Argoudelis, A.D.; Baczynskyj, L.; Mizsak, S.A.; Shilliday, F.B.; Spinelli, P.A.; DeZwaan, J. Paldimycins A and B and antibiotics $273a_{2\alpha}$ and $273a_{2\beta}$ synthesis and characterization. *J. Antibiot.* **1987**, *40*, 419–436. [CrossRef] [PubMed]

19. Fernández, E.; Weissbach, U.; Sánchez Reillo, C.; Braña, A.F.; Méndez, C.; Rohr, J.; Salas, J.A. Identification of two genes from *Streptomyces argillaceus* encoding two glycosyltransferases involved in the transfer of a disaccharide during the biosynthesis of the antitrumor drug mithramycin. *J. Bacteriol.* **1998**, *180*, 4929–4937. [PubMed]

20. Tamura, K.; Stecher, G.; Peterson, D.; Filipski, A.; Kumar, S. MEGA6: Molecular Evolutionary Genetics Analysis version 6.0. *Mol. Biol. Evol.* **2013**, *30*, 2725–2729. [CrossRef] [PubMed]

21. Sievers, F.; Wilm, A.; Dineen, D.G.; Gibson, T.J.; Karplus, K.; Li, W.; Lopez, R.; McWilliam, H.; Remmert, M.; Söding, J.; et al. Fast, scalable generation of high-quality protein multiple sequence alignments using Clustal Omega. *Mol. Syst. Biol.* **2011**, *7*. [CrossRef] [PubMed]

22. Kimura, M. A simple method for estimating evolutionary rates of base substitutions through comparative studies of nucleotide sequences. *J. Mol. Evol.* **1980**, *16*, 111–120. [CrossRef] [PubMed]

23. Saitou, N.; Nei, M. The neighbour-joining method: A new method for reconstructing phylogenetic trees. *Mol. Biol. Evol.* **1987**, *4*, 406–425. [PubMed]

24. Felsenstein, J. Conference limits on phylogenies: An approach using the bootstrap. *Evolution* **1985**, *39*, 783–791. [CrossRef] [PubMed]

25. Audoin, C.; Bonhomme, D.; Ivanisevic, J.; de la Cruz, M.; Cautain, B.; Monteiro, M.C.; Reyes, F.; Rios, L.; Perez, T.; Thomas, O.P. Balibalosides, an Original Family of Glucosylated Sesterterpenes Produced by the Mediterranean Sponge *Oscarella balibaloi*. *Mar. Drugs* **2013**, *11*, 1477–1489. [CrossRef] [PubMed]

marine drugs

MDPI

Article

The Marine Natural Product Pseudopterosin Blocks Cytokine Release of Triple-Negative Breast Cancer and Monocytic Leukemia Cells by Inhibiting NF-κB Signaling

Julia Sperlich [1], Russell Kerr [2] and Nicole Teusch [1,*]

[1] Bio-Pharmaceutical Chemistry & Molecular Pharmacology, Faculty of Applied Natural Sciences, Technische Hochschule Koeln, Chempark, 51368 Leverkusen, Germany; julia.sperlich@th-koeln.de

[2] Department of Chemistry, and Department of Biomedical Sciences, Atlantic Veterinary College, University of Prince Edward Island, Charlottetown, PE C1A 4P3, Canada; rkerr@upei.ca

* Correspondence: nicole.teusch@th-koeln.de; Tel.: +49-214-32831 (ext. 4623)

Received: 16 May 2017; Accepted: 21 August 2017; Published: 23 August 2017

Abstract: Pseudopterosins are a group of marine diterpene glycosides which possess an array of biological activities including anti-inflammatory effects. However, despite the striking in vivo anti-inflammatory potential, the underlying in vitro molecular mode of action remains elusive. To date, few studies have examined pseudopterosin effects on cancer cells. However, to our knowledge, no studies have explored their ability to block cytokine release in breast cancer cells and the respective bidirectional communication with associated immune cells. The present work demonstrates that pseudopterosins have the ability to block the key inflammatory signaling pathway nuclear factor κB (NF-κB) by inhibiting the phosphorylation of p65 and IκB (nuclear factor of kappa light polypeptide gene enhancer in B-cells inhibitor) in leukemia and in breast cancer cells, respectively. Blockade of NF-κB leads to subsequent reduction of the production of the pro-inflammatory cytokines interleukin-6 (IL-6), tumor necrosis factor alpha (TNFα) and monocyte chemotactic protein 1 (MCP-1). Furthermore, pseudopterosin treatment reduces cytokine expression induced by conditioned media in both cell lines investigated. Interestingly, the presence of pseudopterosins induces a nuclear translocation of the glucocorticoid receptor. When knocking down the glucocorticoid receptor, the natural product loses the ability to block cytokine expression. Thus, we hypothesize that pseudopterosins inhibit NF-κB through activation of the glucocorticoid receptor in triple negative breast cancer.

Keywords: pseudopterosin; NF-κB; p65; inflammation; tumor microenvironment; breast cancer; cytokine release; IL-6; TNFα; MCP-1; glucocorticoid receptor

---

1. Introduction

Cancer represents one of the diseases with the highest unmet medical need, causing the second highest incidence of death after cardiovascular diseases in industrialized countries. Among the different types of malignant tumors, breast cancer is the leading cause of cancer mortalities in women worldwide [1]. Classification of breast cancer subtypes is based on the expression of progesterone receptor (PR), estrogen receptor (ER) and/or human epidermal growth factor receptor (HER2). Accordingly, the breast cancer subtype expressing none of these three receptors, the so-called triple-negative breast cancer (TNBC), represents the most aggressive form with currently no targeted therapy available and a significantly reduced overall survival rate [2,3]. Thus, development of innovative and more effective therapies is urgently needed.

Marine organisms represent a vast source of biologically active compounds with a highly unexploited potential for innovative drug development [4]. For instance, the soft coral *Antillogorgia elisabethae* (formerly *Pseudopterogorgia elisabethae*) has been reported to produce at least 31 different secondary metabolites, most of which have not been pharmacologically unexplored [5]. Amongst others, the pseudopterosin family displays a broad spectrum of biological activities, including anti-inflammatory [6–8], analgesic [6,9,10], wound-healing [7,8] and neuromodulatory [11] activity. Moreover, pseudopterosins have shown anti-inflammatory efficacy in phase II clinical trials [12,13] and represent the first commercially licensed marine natural product for use in cosmetic skin care [7,11]. Intriguingly, in vivo assays revealed a higher efficacy of pseudopterosins against topically induced inflammation than the marketed drug indomethacin [6]. Despite the striking in vivo pharmacological effect [6,10,14] and the application in cosmetic products [7,11] the underlying in vitro mechanism of action of the anti-inflammatory potential of pseudopterosins remains elusive. The potential of pseudopterosin A (PsA) has been described as spreading across different intracellular mechanisms ranging from inhibition of phospholipase A2 [10], altering calcium release [15], and inducing cytotoxicity in cancer cells [16]. To our knowledge, no studies have explored the potential of pseudopterosins as a novel immune modulatory agent in breast cancer.

A key factor in regulating inflammatory responses is the transcription factor nuclear factor κB (NF-κB) that acts by controlling expression of cytokines and chemokines. Activation can be triggered by various factors including pro-inflammatory cytokines, growth factors, hormones, oxidative stress, viral infections or DNA-damaging agents [17–20]. Pathogen-associated-molecular-patterns (PAMPs) such as lipopolysaccharides (LPS) and tumor necrosis factor alpha (TNFα) are ligands of different receptors, both triggering activation of the NF-κB-controlled immune response [21–23]. The NF-κB family consists of five functionally conserved members in mammalian cells, including RELA (nuclear factor NF-kappa-B subunit p65), RELB (nuclear factor NF-kappa-B subunit p60), c-REL, NF-κB1 (p105 and p50) and NF-κB2 (p100 and p52) [24]. The specific activation of NF-κB in the innate and adaptive immune defense is opposed by constitutive NF-κB expression in various tumor types. Constitutive activation of NF-κB could be confirmed in cancer in general, and in breast cancer in particular, supporting overall tumor progression, drug resistance, invasiveness, epithelial-to-mesenchymal-transition (EMT) and the promotion of hormone-independent growth [17,25–28]. Elevated NF-kB activity has been observed in both primary human breast cancer tissues and breast cancer cell lines. Furthermore, a recent study assigned a key role of NF-kB in disrupting important microenvironmental cues necessary for tissue organization [29]. The tumor microenvironment (TME) encompasses a complex interplay between tumor cells and tumor associated immune cells. Tumor associated macrophages (TAM) play a crucial role in cancer progression [30]. Tumor associated macrophages produce high amounts of cytokines such as interleukin-6 (IL-6), interleukin-8 (IL-8), monocyte chemotactic protein 1 (MCP-1) and tumor necrosis factor alpha (TNFα) to alter the tumor progression in different ways. IL-6 promotes tumor proliferation, IL-8 leads to neovascularization, growth, angiogenesis and metastasis, and TNFα affects necrosis, invasion and metastasis [26,27]. Moreover, MCP-1 overexpression correlates with histological grade and low level differentiation in breast tumors [31].

The glucocorticoid receptor alpha (GR) has been investigated in different clinical studies as a putative pharmacological target for the treatment of breast cancer [32–34]. Interestingly, there is evidence that NF-κB and GR can physically interact and heterodimerize in breast cancer [35]. By binding other transcription factors such as NF-κB or AP-1, GR can either transactivate or suppress its target genes [1]. Agonism of glucocorticoids (GC) can block migration, invasion and angiogenesis via down-regulation of IL-6 and IL-8 and has been reported to induce drug sensitivity. Furthermore, GC activation induces apoptosis in lymphoid cancer and MCF-7 breast cancer cells [36–38]. However, due to high variability in its expression frequency, divergent cellular functions of GR have been described [2]. Herein, we describe inhibitory capabilities of a mixture of pseudopterosins on the NF-κB signaling pathway and its target genes, the cytokines, in monocytic leukemia and in triple negative breast cancer cells (TNBC) presumably by agonizing the glucocorticoid receptor α. Moreover, our study ascribes the efficient cytokine blockade in the context of bidirectional tumor-immune-cell communication to pseudopterosin treatment.

2. Results

2.1. Pseudopterosin Reduces Cytokine Release by Inhibition of NF-κB Signaling

Pseudopterosins have been described as anti-inflammatory agents with an unknown in vitro mechanism of action. To explore intracellular signaling pathways following pseudopterosin treatment, we investigated the influence of an extract mixture containing four different pseudopterosin derivatives (PsA-D) on the key inflammatory signaling pathway NF-κB. For this purpose, we generated a stable cell line based on the triple negative breast cancer cell line MDA-MB-231 (subsequently named NF-κB-MDA-MB-231) (see Section 4.2, Stable Cell Line Generation). MDA-MB-231 cells display a high level of toll-like-receptor 4 (TLR4) [39] which can activate NF-κB signaling via its ligand LPS [40]. Interestingly, increasing amounts of pseudopterosin inhibited LPS-induced NF-κB activation in NF-κB-MDA-MB-231 breast cancer cells in a concentration-dependent manner (Figure 1A) with an IC$_{50}$ value of 24.4 μM. Additional studies revealed that pseudopterosin also reduced NF-κB activation initiated by other stimuli including TNFα (Figure S1). Moreover, addition of 30 μM of pseudopterosin in monocytic THP-1 cells led to a 1.65-fold inhibition of NF-κB-dependent luciferase activity (Figure 1B).

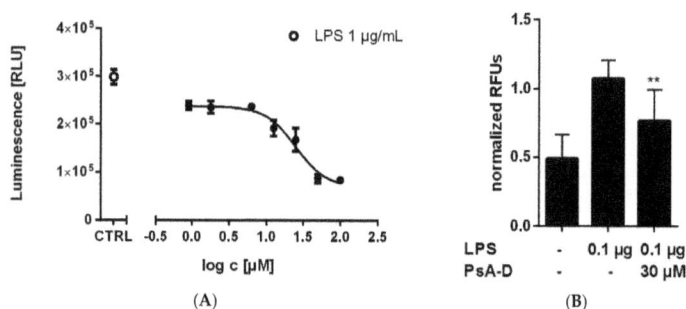

Figure 1. Nuclear factor κB (NF-κB) inhibition in lipopolysaccharide (LPS)-stimulated stable NF-κB-MDA-MB-231 and THP-1 monocytic leukemia cells. (**A**) Dose–response curve of pseudopterosin (PsA-D) on LPS stimulated NF-κB-MDA-MB-231 cells expressing a luciferase reporter gene which is under the control of a NF-κB CMV (cytomegalovirus) promotor. Luminescence intensity correlates proportionally with NF-κB activation. The solid circle represents NF-κB induction in the presence of 1 μg/mL LPS (positive control). PsA-D treatment was performed for 20 min in a bisecting titration followed by 1 μg/mL LPS for 1 h. IC$_{50}$ value of 24.4 μM of pseudopterosin was calculated from three independent experiments; (**B**) Inhibition of NF-κB upon pseudopterosin treatment in THP-1 monocytic leukemia cells (ELISA). Cells were incubated with PsA-D for 20 min followed by LPS treatment. Pseudopterosin decreased NF-κB activation significantly. RLU = relative luminescence units; RFU = relative fluorescence units. Two stars represent a significance of $p < 0.05$. Error bars were calculated using standard error of the mean (+SEM); $n = 3$.

As multiple pro-inflammatory cytokines such as IL-1, IL-6 and TNFα represent target genes of NF-κB [41–43], we investigated the effect of PsA-D on pro-inflammatory cytokine release.

Analyzing a subset of six different cytokines simultaneously, in THP-1 cells incubated with 1 μg/mL LPS led to a significant secretion of IL-6, TNFα and MCP-1 compared with unstimulated control (23-fold induction of IL-6, 33-fold induction of TNFα and 24-fold increase of MCP-1; Table 1), but not IL-1β, IL-12 or IL-4 (data not shown). Compared to THP-1 cells, MDA-MB-231 breast cancer cells displayed a higher basic level of IL-6 and MCP-1. Upon LPS stimulation, we confirmed a 3-fold increase of IL-6, a 15-fold induction of TNFα and a 5-fold increase of MCP- 1 in MDA-MB-231 cells (Table 1). In contrast, no induction of IL-1β or IL-4 could be observed in the triple negative breast cancer cells (data not shown). In both cell lines investigated, PsA-D incubation was able to induce a significant blockade of cytokine secretion: In THP-1 monocytic leukemia cells pseudopterosin reduced TNFα release by at least 47%, blocked IL-6 release by 50% and MCP-1 release by 73%. In MDA-MB-231 breast cancer cells incubated with PsA-D led to a reduction of MCP-1 by 85%, a decrease of TNFα release by 75%, and a decrease of IL-6 by 38%.

Table 1. Inhibition of cytokine release in THP-1 monocytic leukemia and MDA-MB-231 triple negative breast cancer. THP-1 cells were treated with 10 ng PMA (phorbol 12-myristate 13-acetate) for 24 h to induce differentiation. Cytokine amounts were analyzed in supernatants after a 24-h incubation time. Total amounts of cytokines (pg/mL) were calculated according to a standard concentration curve. No treatment serves as a control. % inhibition reflects the percentage of cytokines reduced by PsA-D treatment. Standard deviation was calculated for amounts of cytokines (±SD); $n = 3$. TNF: tumor necrosis factor alpha; IL: interleukin; MCP: monocyte chemotactic protein 1.

MDA-MB-231	Control (pg/mL)	+LPS 1 µg/mL	+PsA-D 30 µM	*p*-Value	% Inhibition
IL-6	1626.3 ± 144	4666.7 ± 307	2874.8 ± 610	<0.0002	38.3
TNFα	1.9 ± 0.6	29.1 ± 5.5	7.17 ± 3.4	<0.0005	75.3
MCP-1	325.3 ± 260	1625.6 ± 540.6	241.3 ± 100.9	0.0082	85.2
THP-1	**Control (pg/mL)**	**+LPS 1 µg/mL**	**+PsA-D 30 µM**	***p*-Value**	**% Inhibition**
IL-6	2.8 ± 1	66.7 ± 9.8	33 ± 1.98	0.0089	50.0
TNFα	13.4 ± 4.5	439.4 ± 28	232.0 ± 100	0.1138	47.2
MCP-1	182.9 ± 65.3	4436.7 ± 2098	1208.9 ± 762.3	0.0552	72.8

As the NF-κB signaling pathway can be activated with different stimuli including LPS, TNFα or pathogen-associated molecular patterns (PAMPs) [18,44,45], we utilized TNFα, the ligand of the TNFα receptor 1 (TNFR1) [23,46], to induce NF-κB signaling independent of TLR4. As expected, stimulation with TNFα increased the expression levels of the investigated cytokines in MDA-MB-231 breast cancer cells significantly compared to unstimulated control (IL-6 4-fold, IL-8 6-fold, MCP-1 5-fold) (Figure 2A). It is noteworthy that pseudopterosin blocked the expression of all cytokines investigated, however, statistical significance was only noted for IL-6 and MCP-1 (IL-6 2.7-fold induction, MCP-1 3.7-fold induction).

Secretion of cytokines is stimulated after TNFα treatment (IL-6 4540 ± 329 pg/mL, IL-8 4047 ± 196 pg/mL, MCP-1 4048 ± 18 pg/mL) (Figure 2B). Cytokine amounts declined in the triple negative breast cancer cells in a concentration-dependent manner upon pseudopterosin treatment (at a PsA-D concentration of 30 µM: 18-fold decrease of IL-6, 12-fold reduction of IL-8 and a 26-fold decrease of MCP-1). Significant inhibition at a concentration of 10 µM of PsA-D could be achieved for MCP-1 (6-fold decrease of MCP-1 release compared to untreated control).

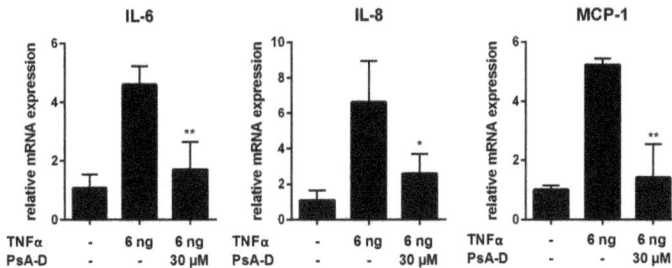

(A)

Figure 2. *Cont.*

(B)

(C)

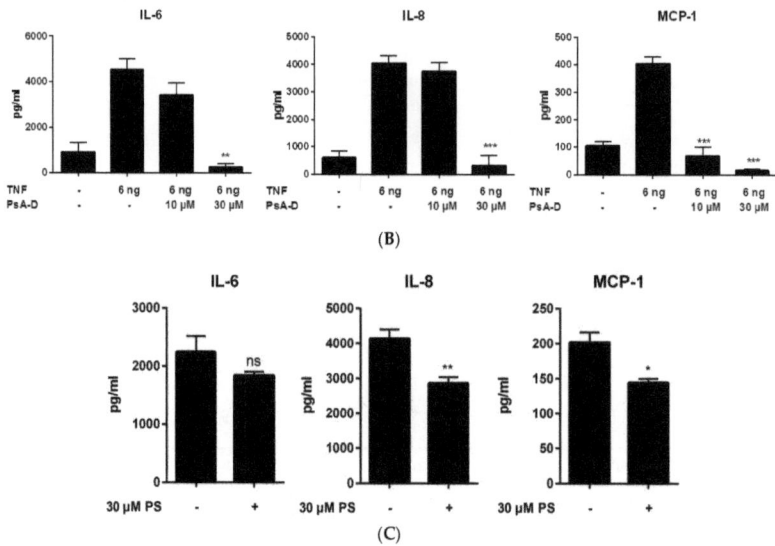

Figure 2. Inhibition of cytokine expression (**A**) and secretion (**B**) after TNFα stimulation and inhibition of endogenous cytokine secretion (**C**) in MDA-MB-231 triple-negative breast cancer (TNBC). (**A**) MDA-MB-231 cells were treated with 30 μM of PsA-D for 20 min followed by 6 ng/mL of TNFα for 5 h; (**B**) Various concentrations of PsA-D were incubated for 20 min followed by TNFα treatment for 24 h; (**C**) MDA-MB-231 cells were treated with 30 μM of PsA-D and cytokine secretion was measured 24 h thereafter. Error bars were calculated using +SEM; $n = 3$. p-Values are calculated against TNFα. Three stars represent a significance of $p < 0.001$, two stars $p < 0.01$, one star $p < 0.05$ and "ns." means not significant.

It is noteworthy that irrespective of exogenous cytokine stimulation via LPS or TNFα, pseudopterosins are able to significantly reduce endogenous release of at least two cytokines in the MDA-MB-231 triple negative breast cancer cells (IL-6 1.2-fold, IL-8 1.4-fold, MCP-1 1.4-fold) (Figure 2C). Moreover, additional investigation demonstrates that the reported inhibitory effect of PsA-D on cytokine release can be assigned to other triple negative cell lines (Table S1).

2.2. Pseudopterosin Blocks Bidirectional Communication

To explore whether pseudopterosins have the ability to inhibit the bidirectional communication between immune cells and tumor cells, we designed an experimental set-up imitating inter-cell communication within the tumor microenvironment (Figure 3A). As shown, stimulation by LPS leads to the production of secondary metabolites including cytokines and the subsequent secretion into the surrounding "conditioned medium" (CM). Medium containing cytokines released by MDA-MB-231 cells represents the so called "MDA-MB-231 conditioned medium" (M-CM; Figure 3B), whereas medium encompassing cytokines secreted by THP-1 cells referred to as "THP-1 conditioned medium" (THP-CM; Figure 3C). Both conditioned media were used in independent experiments to stimulate the respective opposite cell line. Treatment with unstimulated conditioned medium did not influence cytokine expression in any of the investigated cell lines. However, incubation of THP-1 leukemia cells with stimulated M-CM induced a significant cytokine expression in THP-1 cells (8-fold increase of IL-6, 18-fold induction of TNFα and nearly 13-fold in MCP-1 expression). Furthermore, the triple negative breast cancer cell line MDA-MB-231 induced expression of IL-6, TNFα and MCP-1 in the presence of stimulated THP-CM (IL-6 induction 177-fold, TNFα induction nearly 10-fold and MCP-1 induction nearly 19-fold).

Notably, pseudopterosin treatment was able to block cytokine expression induced by conditioned media in both leukemia cells and in triple negative breast cancer cells. In THP-1 cells stimulated with M-CM, a 2-fold reduction of IL-6 expression and a 3-fold reduction of MCP-1 expression was noted following pseudopterosin treatment. Also, MDA-MB-231 cells stimulated with THP-CM displayed a 4-fold increase in IL-6 and a 2.5-fold increase in MCP-1 expression. In conclusion, our data demonstrate that PsA-D is able to significantly decrease expression of the cytokines IL-6 and MCP-1 after stimulation with pre-conditioned medium in monocytes and breast cancer cells, respectively.

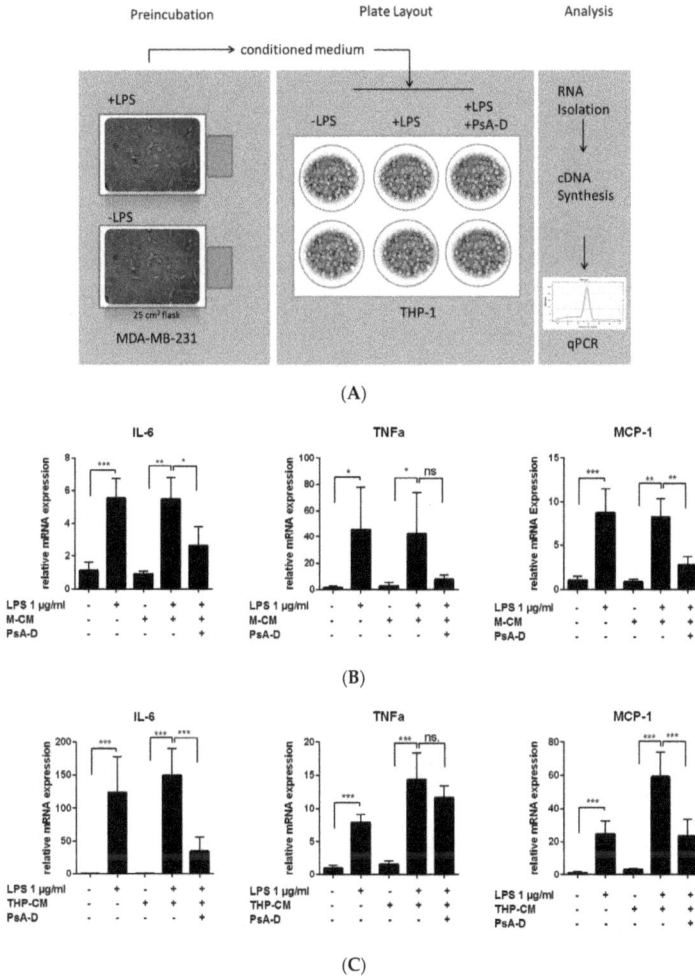

(A)

(B)

(C)

Figure 3. Blockage of bidirectional communication between THP-1 monocytic leukemia and MDA-MB-231 TNBC. (**A**) Process scheme of producing tumor conditioned medium. THP-1 or MDA-MB-231 cells were cultured in 25 cm^2 flasks and treated with 1 µg/mL LPS for 24 h. Medium was collected and centrifuged. After sterile filtration, tumor conditioned medium was added to seeded cells in 6-well plates. (**B**) MDA-MB-231 conditioned medium (M-CM) or (**C**) THP-1 conditioned medium (THP-CM) was added to the adversary cells. RNA was isolated for further analysis in real-time PCR. Error bars were calculated using +SEM. *p*-Values of three stars represent a significance of *p* < 0.001, two stars *p* < 0.01, one star *p* < 0.05 and "ns." means not significant.

To exclusively ascribe the demonstrated cytokine expression patterns to the pre-treatment with the respective conditioned medium, we subjected MDA-MB-231 cells to a knock-down of the TLR4 receptor (siRNA-TLR4 (siTLR4) transfected cells) (Figure 4A). As a control, we transfected non-coding silencing RNA (nc siRNA). A 50% TLR4 knock down was achieved. Compared to a nc siRNA control, siTLR4 transfection did not influence TNFα expression level upon pseudopterosin treatment. Monitoring the p65 phosphorylation with TNFα and LPS in parallel experiments we confirmed a 2-fold reduction of phosphorylation after pseudopterosin treatment independent of the stimulus (Figure 4B). In conclusion, PsA-D induced cytokine blockade and p65 phosphorylation in triple negative breast cancer cells does not dependent on TLR4.

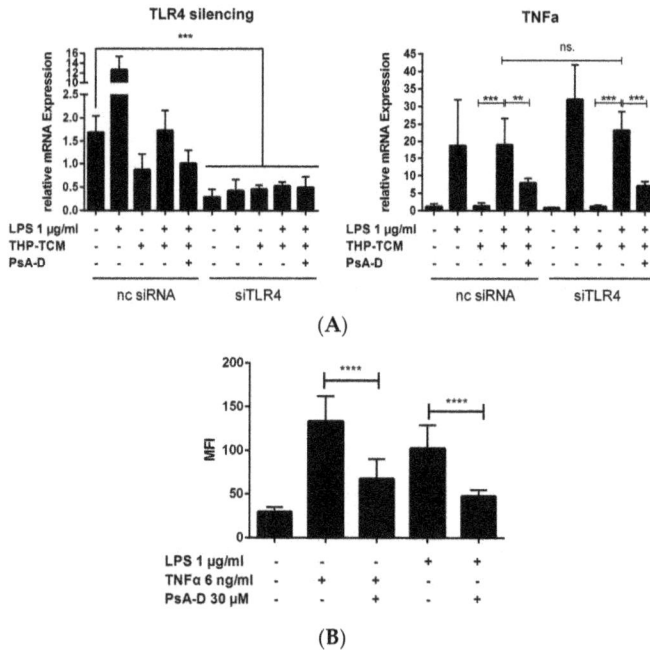

(A)

(B)

Figure 4. PsA-D induced NF-κB inhibition is toll-like-receptor-4 (TLR4)-independent. (A) MDA-MB-231 cells were seeded in 6-well plates and incubated for 24 h. Transfection with 2 μM siRNA was done with Lipofectamine3000 following the manufacturer's protocol. After 24 h, cells were first treated with 30 μM PsA-D before and following treatment with THP-CM for 5 h. After another 24 h of incubation, cells were harvested and lysed for RNA isolation in preparation for realtime PCR. Knock-down efficiency of TLR4 was about 50%. PsA-D blocked TNFα expression independent of TLR4 expression; (B) MDA-MB-231 cells were stimulated either with 1 μg/mL LPS or with 6 ng/mL TNFα following 20 min treatment of PsA-D. P65 phosphorylation was measured after 24 h of treatment. Error bars were calculated using +SEM. *p*-Values of four stars show a significance of $p < 0.0001$, three stars $p < 0.001$, two stars $p < 0.01$ and "ns." means not significant.

2.3. Pseudopterosin Inhibits NF-κB through Activation of the Glucocorticoid Receptor

Our data show for the first time that the underlying in vitro mechanism of the well described anti-inflammatory response of pseudopterosin might be ascribed to inhibition of the NF-κB pathway. To further explore putative molecular pharmacological targets of pseudopterosins, we started to investigate the influence of the natural product on glucocorticoid signaling. NF-κB and glucocorticoid receptor α (GR) display opposed functions in regulating immune and inflammatory responses. Moreover, both transcription factors have been described as transcriptional antagonists [36]. Thus, we investigated the interaction of pseudopterosin with GR. To evaluate transactivation of GR in the presence of PsA-D on the whole cell level, we used immunofluorescent staining of GR in MDA-MB-231 cells incubating the cells with dexamethasone, serving as a positive control, or PsA-D (Figure 5A). Untreated cells displayed an even GR distribution within the cytosol, whereas the nucleus did not show any GR localization. As expected, upon dexamethasone treatment the GR staining revealed a complete translocation of the receptor to the nucleus in breast cancer cells. Interestingly, the presence of pseudopterosin induced a comparable nuclear translocation of the GR. Quantification of the respective fluorescence intensities using the software ImageJ confirmed a significant GR translocation to the nucleus after dexamethasone treatment (4.5-fold reduction of cytoplasmic total corrected cell fluorescence (TCCF) compared to control) and pseudopterosin treatment (2.5-fold reduction of cytoplasmic total corrected cell fluorescence (TCCF) compared to control, Figure 5B). Accordingly, PsA-D inhibited phosphorylation of p65 and $I\kappa B\alpha$ significantly compared to LPS stimulation (Figure 5C) or compared to stimulation with $TNF\alpha$ (Figure S2) (2-fold inhibition, respectively).

Moreover, to confirm GR as a putative pharmacological target of pseudopterosin we performed a glucocorticoid receptor α knock-down in MDA-MB-231 cells. In this context, we transfected cells with siRNA of GR (siGR, Figure 6) with non-coding siRNA (nc siRNA) serving as a negative control. A 60% knock-down of GR was achieved. Treatment with negative control nc siRNA revealed that unaltered GR expression resulted in cytokine expression level after LPS stimulation comparable to previous results (Table 1). Furthermore, as demonstrated earlier, pseudopterosin inhibited IL-6 (3-fold) and MCP-1 (nearly 4-fold) significantly in the presence of GR. However, when knocking down GR, pseudopterosin lost the ability to block IL-6 or MCP-1 expression, respectively. To finally confirm glucocorticoid receptor α as a potential pharmaceutical target for pseudopterosin, we used a reporter gene assay expressing a luciferase under the control of a human GR promotor (Figure 6B). In line with our previous findings, pseudopterosin induced a significant increase in expression of human GR. In conclusion, the described inhibitory effect of pseudopterosin on cytokine expression and release in triple negative breast cancer is putatively ascribed to agonism of glucocorticoid receptor α.

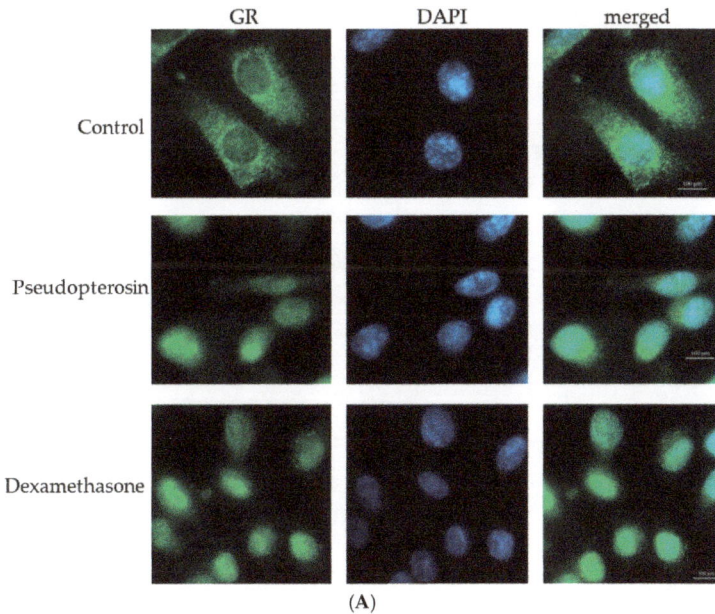

(A)

Figure 5. *Cont.*

(B)   (C)

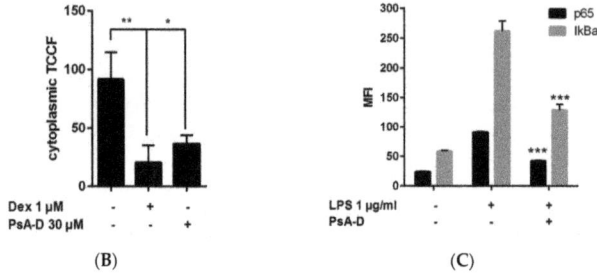

Figure 5. Pseudopterosin-induced activation of glucocorticoid receptor alpha (GR) translocation into the nucleus is accompanied by inhibition of phosphorylation of p65. (**A**) PsA-D was added at a concentration of 30 µM and dexamethasone at 1 µM in MDA-MB-231 cells. Cell nuclei were stained with 3 µM 4′,6-Diamidin-2-phenylindol (DAPI; blue channel). GR is shown in green. The right column shows merged channels; (**B**) Quantification of immunofluorescence staining shows cytoplasmic total corrected cell fluorescence (TCCF). TCCF was calculated as described in methods section. Cytoplasmic TCCF was calculated after following formula: TCCF GFP–TCCF DAPI. Cytoplasmic staining reduced significantly after dexamethasone (Dex) or PsA-D treatment; (**C**) Phosphorylation of p65 and IκBα induced by LPS was investigated in the absence or presence of PsA-D with an incubation time of 20 min on MDA-MB-231 breast cancer cells. *p*-Values of three stars show a significance of $p < 0.001$, two stars of $p < 0.01$ and one star of $p < 0.05$; +SEM; $n = 30$. MFI = median fluorescence intensity.

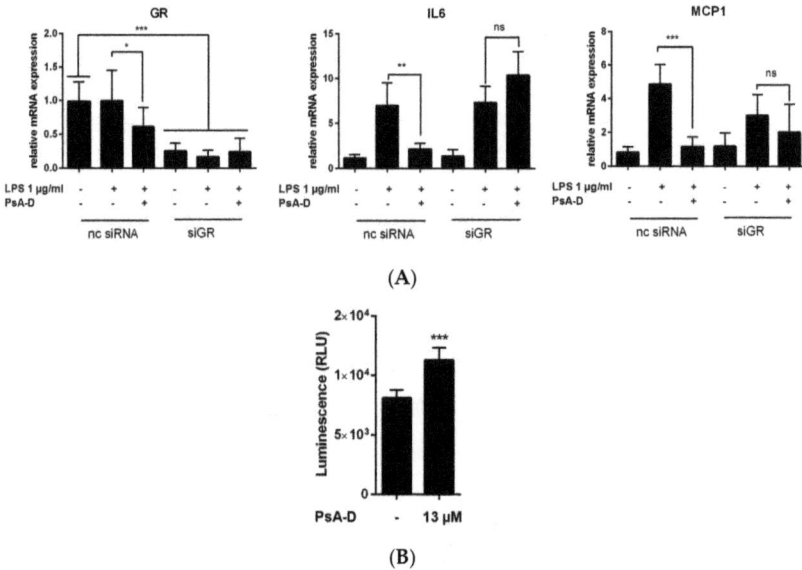

(A)

(B)

Figure 6. Pseudopterosin as a low molecular weight agonist of GR. (**A**) MDA-MB-231 cells were seeded in 6-well plates and transfected with 2 µM siRNA with the Nucleofector® 2b device using the manufacturer's protocol. After 24 h, cells first were treated with 30 µM PsA-D for 20 min and subsequently with 1 µg/mL LPS for 24 h. After another 24 h of incubation, cells were harvested and lysed for RNA isolation as preparation for further real-time PCR analysis; (**B**) Cells were seeded following manufacturer's instructions. Reporter cells stably expressing a luciferase under the control of a human GR promotor were activated upon pseudopterosin treatment. Error bars were calculated using +SEM; (**A**) $n = 3$; (**B**) $n = 2$. *p*-Values of three stars show a significance of $p < 0.001$, two stars $p < 0.01$, one star $p < 0.05$ and "ns." means not significant.

3. Discussion

Though their mechanism of action remains unknown, pseudopterosins have been demonstrated as anti-inflammatory [6–8], analgesic [6,9,10], wound-healing [7,8], anti-microbial [47,48], and anti-cancer agents [16]. In our work we were able to illuminate a novel molecular mechanism of the broadly described anti-inflammatory activity of pseudopterosin by demonstrating a concentration-dependent inhibition of the NF-κB pathway based on inhibition of p65 and IκB phosphorylation.

NF-κB overexpression maintains cancer stem cell populations in the basal-subtype of breast cancer and plays a crucial role in overall cancer progression [29,49–51]. NF-κB activity is involved in epithelial-to-mesenchymal transition (EMT) [52]. Thus, previous studies have approached the inhibition of NF-κB activity in several ways: Gordon et al. suppressed NF-κB transcription in MDA-MB-231 breast cancer cells resulting in reduced osteolysis after tumor cell injection in mice combined with decreased cytokine expression [53]. Furthermore, inhibition of NF-κB activity in human breast cancer cells (MDA-MB-231 and HCC1954) reduced invasiveness and migration [52]. In conclusion, NF-κB activation blockade demonstrates effective reduction in tumor growth and progression. Our study revealed pseudopterosin to efficaciously inhibit NF-κB signaling and subsequent cytokine release in both THP-1 monocytic leukemia cells and MDA-MB-231 breast cancer cells. Furthermore, pseudopterosin has demonstrated the ability to block the inter-cell communication between immune cells and MDA-MB-231 breast cancer cells, a complex interplay presumably important within the tumor microenviromental set-up.

Nuclear receptors like the glucocorticoid receptor α (GR) translocate into the nucleus upon activation and bind the glucocorticoid response element (GRE) enabling the transcription of target genes ultimately resulting in immune suppression. Thus, GR and NF-κB are transcription factors with opposing functions in regulating inflammatory responses. In cancer therapy glucocorticoids are used as a pre-treatment combined with chemotherapy to prevent vomiting and allergic reactions [32,38,54]. However, due to high variability in its expression frequency, divergent cellular functions of GR have been described [2]. For instance, high expression levels not only lead to poor prognosis for ER⁻ breast cancer patients, but are also associated with better outcomes in patients with ER⁺ breast cancer [55]. Suppression of chemotherapy induced apoptosis for example is correlated with high GR expression and poor prognosis [37,55,56]. On the other hand, glucocorticoids can suppress migration, invasion and angiogenesis via down-regulation of IL-6 and IL-8. Furthermore, GR agonism has been shown to induce drug sensitivity and apoptosis in lymphoid cancer and breast cancer [36–38].

Interestingly, there is evidence that expression of both transcription factors, NF-κB and GR, are correlated in the context of breast cancer. While NF-κB is up-regulated [25,57], GR over-expression could be confirmed for breast cancer, however, in contrast to NF-κB, GR levels decreased during cancer progression [58]. Furthermore, there is evidence that NF-κB and GR can even physically interact by heterodimerization [35,51]. Glucocorticoids regulate target genes by either positive or negative regulatory mechanisms. Anti-inflammatory effects are mediated via a transcription repressive function (so called transrepressive action) of GR, whereas activation of gene transcription (namely transactivation) results in an undesirable side effect of glucocorticoids including chemoresistance, impaired wound-healing, and skin and muscle atrophy [59–61]. A previous study revealed that NF-κB inhibition is likely based on the transrepressive function of GR [1]. Our study confirms GR as putative pharmacological target of pseudopterosins. In conclusion, we hypothesize that the induction of GR activation upon pseudopterosin treatment might be based on GR acting as transrepressive on NF-κB.

As triple-negative breast cancer represents one of the diseases with a high unmet medical need resulting in a low overall survival rate, there is a need for efficacious drug treatment regimens. Our study contributes by elucidating the molecular mode of action of the striking anti-inflammatory effect of the marine diterpene glycosides PsA-D in the context of breast cancer. Thus, we demonstrate the mostly unexplored pharmaceutical potential of pseudopterosins as a promising basis for developing novel cancer treatment strategies. Future studies may include a medicinal chemistry approach to design simplified derivatives of pseudopterosin with improved potency.

4. Materials and Methods

4.1. Cell Culture and Commercially Available Reagents

TNFα was purchased from Peprotech (Rocky Hill, NJ, USA). MDA-MB-231 breast cancer cells were obtained from European Collection of Authenticated Cell Cultures (Salisbury, UK) and grown in humidified atmosphere containing no CO_2 in Leibovitz's L15 medium. Medium was supplemented with 15% FCS (fetal calf serum), 2 mM glutamine, 100 unit's mL^{-1} penicillin and 100 µg mL^{-1} unit's streptomycin. THP-1 acute monocytic leukemia cells were purchased from the German Collection of Microorganisms and Cell Culture (Braunschweig, Germany) and cultured in the presence of 5% CO_2 in RPMI along with 10% FCS, penicillin and streptomycin. This cell line was used as a model for cells derived from the immune system. Medium and antibiotics were purchased from Gibco (Life Technologies, Carlsbad, CA, USA).

4.2. Stable Cell Line Generation

MDA-MB-231 breast cancer cells were used to create a stable cell line subsequently named NF-κB-MDA-MB-231 where the expression of a Luciferase reporter gene is under the control of a NF-κB CMV promoter. The vector was purchased from Promega (Madison, WI, USA): pNL3.2.NF-κB-RE[NlucP/NF-κB-RE/Hygro]. Cells were transfected with the nucleofector 2b device from Lonza Group AG (Basel, Switzerland) and the corresponding RCT Cell Line Kit V according to the manufacturer's protocol. Cells were cultured in DMEM supplemented with 10% FCS, 100 units mL^{-1} penicillin and 100 units mL^{-1} streptomycin. After transfection cells were diluted serially to obtain monoclonal cells. After colony formation hygromycin (Sigma, Munich, Germany) clones were cultivated in the presence of hygromycin.

4.3. NF-κB Reportergene Assay

To determine NF-κB activation, cells were seeded with a density of 5×10^5 cells per mL in 384-well plates using the CyBio® pipetting roboter (Analytic Jens AG; Jena, Germany). After 24 h of incubation, cells were treated with different concentrations of PsA-D for 20 min. Afterwards, cells were treated with 1 µg/mL LPS or 6 ng/mL TNFα for 1 h, respectively. Luciferase activity was detected with the NanoGlo Luciferase Assay from Promega. NanoGlo Substrate and buffer were pre-mixed in 1:50 ratio and reagent was added to the wells in a 1:1 ratio and luminescence was determined immediately.

4.4. NF-κB and Human Cytokine Magnetic Bead Kit

MDA-MB-231 breast cancer cells were cultured in 10 cm dishes in 1.8×10^6 cells per mL and incubated for 24 h at 37 °C. Before compound treatment medium was changed to serum-free medium. Cells were treated with PsA-D for 15 min, followed by incubation with 1 µg/mL LPS. Afterwards, cells were lysed with the lysis buffer provided in the NF-κB magnetic bead kit from Merck Millipore (Darmstadt, Germany) to obtain phosphorylated proteins from the nucleus. Protein concentration was determined with Bradford reagent (Roth, Karlsruhe; Germany). Samples were diluted to achieve a concentration of 0.8 mg/mL of total proteins. The subsequent protocol was according to manufacturer's instructions.

MDA-MB-231 breast cancer cells and were seeded in 96-well plates in 4×10^5 cells per mL and MDA-MB-453 in 6×10^5 cells per mL and incubated for 24 h at 37 °C. THP-1 cells were seeded in 4×10^5 cells per mL and after 1 h of incubation differentiated with 10 ng/mL PMA for 24 h. Cells were treated with PsA-D for 20 min and afterwards with 1 µg/mL LPS for 24 h. Supernatant was harvested and stored at -20 °C until measurement of cytokines. The subsequent protocol was performed according to the manufacturer's instructions with the MAGPIX® Multiplexing System from Merck Millipore (Darmstadt, Germany).

4.5. Quantitative Real-Time PCR

To determine cytokine expression levels after PsA-D treatment, the following primers were used (purchased from Eurofins, Ebersberg): IL-6 forward (GGCACTGGCAGAAAACAACC), IL-6 reverse (GCAAGTCTCCTCATTGAATCC) IL-8 forward: (ACTGAGAGTGATTGAGAGTGGAC), IL-8 reverse: (AACCCTCTGCACCCAGTTTTC), TNFα forward: (GCCTGCTGCACTTTGGAGTG), TNFα reverse: (TCGGGGTTCGAGAAGATGAT), MCP-1 forward: (CCCCAGTCACCTGCTGTTAT), MCP-1 reverse: (TGGAATCCTGAACCCACTTC), GAPDH forward: (TGCACCACCAACTGCTTAGC), GAPDH reverse: (GGCATGGACTGTGGTCATGAG), GR forward: (AAAAGAGCAGTGGAAGGACAGCAC) GR reverse: (GGTAGGGGTGAGTTGTGGTAACG). Total RNA was isolated with QIAGEN (Hilden, Germany) RNA Isolation Kit according to manufacturer's instructions and reverse transcriptase PCR were performed with iScript RT cDNAse Kit from BioRad (Munich, Germany). Real-time PCR was conducted with Quantitect SYBR Green from QIAGEN (Hilden, Germany) based on the following protocol: pre-incubation at 95 °C for 900 s, amplification was performed over 45 cycles (95 °C for 15 s, 55 °C for 25 s and 72 °C for 10 s). No-template controls served as negative control. C_T values were calculated according to the $2^{-\Delta\Delta C_T}$ method [62]. Sample values were normalized to the house-keeping gene GAPDH (glyceraldehyde 3-phosphate dehydrogenase).

4.6. Immunofluorecent Staining

MDA-MB-231 breast cancer cells were seeded in 1×10^5 cells per mL and incubated for 24 h. PsA-D or dexamethasone treatment comprised 30 min. Cells were fixed afterwards with −10 °C cold methanol. Cells were made permeable using 0.1% Triton™ X-100. Antibodies were purchased from Santa Cruz Biotechnology (Dallas, TX, USA): primary antibody (sc-8992 GR (H-300)) incubated 1:50 for 24 h overnight at 4 °C and secondary antibody (sc-2012 IgG-FITC (fluorescein isothiocyanate)) was incubated 1:100 for 2.5 h at room temperature. Cells were washed three times with PBS following each incubation step. For staining, cell nuclei 4′,6-Diamidin-2-phenylindol (DAPI, Sigma) was incubated for 5 min at room temperature at a concentration of 3 μM and washed three times with PBS for 5 min.

Quantification of immunofluorescence intensity was achieved with ImageJ (v1.51k). The shape of the cells was outlined and the area, mean gray fluorescence value and integrated density measured. Several background readings were also measured. The "total corrected cellular fluorescence" (=TCCF) was calculated according to following formula: integrated density—(area of selected cell x mean fluorescence of background readings) [63]. Values of GFP staining were subtracted by values of DAPI staining to obtain cytoplasmic TCCF.

4.7. Conditioned Medium (CM) from Tumor Cells

MDA-MB-231 or THP-1 cells were cultured until 70–90% confluency. 1×10^6 cells were counted and transferred into a 25 cm^2 flask. Cells were either stimulated with 1 μg/mL LPS or without LPS as a negative control. Supernatant was collected after 24 h, centrifuged and sterile filtered. Conditioned medium was stored at −80 °C. MDA-MB-231 or THP-1 cells were seeded at 1×10^6 cells per mL in 6-well plates and incubated for 24 h. PsA-D was added at a concentration of 30 μM for 20 min followed by 25 volume percentage of tumor-conditioned medium for 5 h. Cells were then harvested and RNA isolated for further analysis in real-time PCR.

4.8. Knock-Down Studies

TLR4 siRNA s14194 and Silencer® Select Negative Control No. 2 siRNA was purchased from Life Technologies (Darmstadt, Germany). Glucocorticoid receptor (GR) siRNA was purchased from Santa Cruz Biotechnology (Dallas, TX, USA). SiRNA transfection (2 μM of siRNA) was performed using Lipofectamine3000 from Invitrogen (Carlsbad, CA, USA) according to manufacturer's protocol.

4.9. GR Reportergene Assay

Reportergene assay based on non-human stable cells containing constitutive high-level expression of full-length human GR (NR3C1) were purchased from Indigo Biosciences (State College, PA, USA). Assay was performed according to manufacturer's instructions. PsA-D was added to cells according to the agonist assay described in the protocol and incubated for 24 h at 37 °C.

4.10. Preparation of PsA-D Mixture

A. elisabethae was collected from South Bimini Island, The Bahamas, was dried and extracted in EtOAc/MeOH (1:1) for 48 h. The crude extract was subjected to silica gel chromatography eluting with hexanes and EtOAc to afford a mixture of PsA-D. The ratio was determined to be 85:5:5:5 (PsA:B:C:D) by LC-MS analysis.

4.11. Statistical Analysis

Obtained data represent at least three independent experiments. Error bars show +SEM of the means of triplicate values. Statistical analysis was calculated using one-way-ANOVA followed by Dunnett's multiple comparisons test. When groups were compared with a control and/or comparison of mean values of only two groups, an unpaired student's *t*-Test was applied. $p < 0.05$ was chosen to define statistically significant difference. Figures and data analysis were generated with Graphpad Prism v. 6.07 (Graphpad Software, San Diego, CA, USA).

Supplementary Materials: The following are available online at www.mdpi.com/1660-3397/15/9/262/s1, Figure S1: Pseudopterosin inhibits activation of NF-κB after two different stimuli, Figure S2: Pseudopterosin blocked phosphorylation of p65 and IkBα after TNFα stimulation; Table S1: Inhibition of cytokine release in MDA-MB-453 triple negative breast cancer cells.

Acknowledgments: The research project is financed by a grant of Nicole Teusch provided by the Ministry of Innovation, Science and Research of the federal state of North Rhine-Westphalia, Germany. Furthermore, the Ph.D. training of Julia Sperlich is financed by the graduate program in Pharmacology and Experimental Therapeutics at the University of Cologne which is financially and scientifically supported by Bayer. Russel Kerr would like to thank the Government of The Bahamas for permission to collect *Antillogorgia elisabethae*.

Author Contributions: Nicole Teusch and Julia Sperlich developed the scientific concept and designed the experiments. Julia Sperlich performed the experiments and analyzed the data. Russell Kerr provided the pseudopterosin extract and reviewed the manuscript. Nicole Teusch and Julia Sperlich wrote the manuscript.

Conflicts of Interest: The authors declare no conflict of interest.

References

1. Moutsatsou, P.; Papavassiliou, A.G. The glucocorticoid receptor signalling in breast cancer. *J. Cell. Mol. Med.* **2008**, *12*, 145–163. [CrossRef] [PubMed]
2. Belova, L.; Delgado, B.; Kocherginsky, M.; Melhem, A.; Olopade, O.; Conzen, S. Glucocorticoid receptor expression in breast cancer associates with older patient age. *Breast Cancer Res. Treat.* **2015**, *116*, 441–447. [CrossRef] [PubMed]
3. Pal, S.K.; Childs, B.H.; Pegram, M. Triple negative breast cancer: Unmet medical needs. *Breast Cancer Res. Treat.* **2011**, *125*, 627–636. [CrossRef] [PubMed]
4. Abad, M.J.; Bermejo, P. Bioactive natural products from marine sources. *Stud. Nat. Prod. Chem.* **2001**, *25*, 683–755.
5. Berrué, F.; McCulloch, M.W.B.; Kerr, R.G. Marine diterpene glycosides. *Bioorg. Med. Chem.* **2011**, *19*, 6702–6719. [CrossRef] [PubMed]
6. Mayer, A.M.S.; Jacobson, P.B.; Fenical, W.; Jacobs, R.S.; Glaser, K.B. Pharmacological characterization of the pseudopterosins: Novel anti-inflammatory natural products isolated from the Caribbean soft coral, *Pseudopterogorgia elisabethae*. *Life Sci.* **1998**, *62*, 401–407. [CrossRef]
7. Correa, H.; Valenzuela, A.L.; Ospina, L.F.; Duque, C. Anti-inflammatory effects of the gorgonian *Pseudopterogorgia elisabethae* collected at the Islands of Providencia and San Andrés (SW Caribbean). *J. Inflamm. Lond.* **2009**, *6*. [CrossRef] [PubMed]

8.   Ata, A.; Kerr, R.G.; Moya, C.E.; Jacobs, R.S. Identification of anti-inflammatory diterpenes from the marine gorgonian *Pseudopterogorgia elisabethae*. *Tetrahedron* **2003**, *59*, 4215–4222. [CrossRef]
9.   Look, S.A.; Fenical, W.; Matsumoto, G.K.; Clardy, J. The pseudopterosins: A new class of antiinflammatory and analgesic diterpene pentosides from the marine sea whip *Pseudopterogorgia elisabethae* (Octocorallia). *J. Org. Chem.* **1986**, *51*, 5140–5145. [CrossRef]
10.   Look, S.A.; Fenical, W.; Jacobs, R.S.; Clardy, J. The pseudopterosins: Anti-inflammatory and analgesic natural products from the sea whip *Pseudopterogorgia elisabethae*. *Proc. Natl. Acad. Sci. USA* **1986**, *83*, 6238–6240. [CrossRef] [PubMed]
11.   Caplan, S.L.; Zheng, B.; Dawson-Scully, K.; White, C.A.; West, L.M. Pseudopterosin A: Protection of synaptic function and potential as a neuromodulatory agent. *Mar. Drugs* **2016**, *14*, 55. [CrossRef] [PubMed]
12.   Newman, D.J.; Cragg, G.M. Marine natural products and related compounds in clinical and advanced preclinical trials. *J. Nat. Prod.* **2004**, *67*, 1216–1238. [CrossRef] [PubMed]
13.   Mayer, A.M.S.; Glaser, K.B.; Cuevas, C.; Jacobs, R.S.; Kem, W.; Little, R.D.; Mcintosh, J.M.; Newman, D.J.; Potts, B.C.; Shuster, D.E. The odyssey of marine pharmaceuticals: A current pipeline perspective. *Trends Pharmacol. Sci.* **2010**, *31*, 255–265. [CrossRef] [PubMed]
14.   Mayer, A.M.S.; Rodriguez, A.D.; Berlinck, R.G.S.; Hamann, M.T. Marine pharmacology in 2003–4: Marine compounds with anthelminthic, antibacterial, anticoagulant, antifungal, anti-inflammatory, antimalarial, antiplatelet, antiprotozoal, antituberculosis, and antiviral activities; affecting the cardiovascular, immune. *Comp. Biochem. Physiol. C Toxicol. Pharmakol.* **2008**, *145*, 553–581. [CrossRef] [PubMed]
15.   Moya, C.E.; Jacobs, R.S. Pseudopterosin A inhibits phagocytosis and alters intracellular calcium turnover in a pertussis toxin sensitive site in *Tetrahymena thermophila*. *Comp. Biochem. Physiol. C Toxicol. Pharmacol.* **2006**, *143*, 436–443. [CrossRef] [PubMed]
16.   Rodríguez, I.I.; Shi, Y.P.; García, O.J.; Rodríguez, A.D.; Mayer, A.M.S.; Sánchez, J.A.; Ortega-Barria, E.; González, J. New pseudopterosin and seco-pseudopterosin diterpene glycosides from two Colombian isolates of *Pseudopterogorgia elisabethae* and their diverse biological activities. *J. Nat. Prod.* **2004**, *67*, 1672–1680. [CrossRef] [PubMed]
17.   Badr, C.; Niers, J.M.; Tjon-Kon-Fat, L.-A.; Noske, D.P.; Wurdinger, T.; Tannous, B. Real-time monitoring of NF-kappaB activity in cultured cells and in animal models. *Mol. Imaging* **2009**, *8*, 278–290. [PubMed]
18.   Kawai, T.; Akira, S. The role of pattern-recognition receptors in innate immunity: Update on Toll-like receptors. *Nat. Immunol.* **2010**, *11*, 373–384. [CrossRef] [PubMed]
19.   Ramage, L. Expression of Pro-Inflammatory Proteins in the Lung Epithelial Cell Line A549, in Response to Cytokine, and Environmental Particle Exposure. Ph.D. Thesis, Edinburgh Napier University, Edinburgh, UK, 2003.
20.   Blank, V.; Kourilsky, P.; Israel, A.; Publishers, E.S. NF-kB and related proteins: Rel/dorsal homologies meet ankyrin-like repeats. *Trends Biochem. Sci.* **1992**, *17*, 135–140. [CrossRef]
21.   Baeuerle, P.A.; Baltimore, D. Activation of DNA-binding activity in an apparently cytoplasmic precursor of the NF-kappa B transcription factor. *Cell* **1988**, *53*, 211–217. [CrossRef]
22.   Balkwill, F. TNF-alpha in promotion and progression of cancer. *Cancer Metastasis Rev.* **2006**, *25*, 409–416. [CrossRef] [PubMed]
23.   Zhang, L.; Blackwell, K.; Altaeva, A.; Shi, Z.; Habelhah, H. TRAF2 phosphorylation promotes NF-κB-dependent gene expression and inhibits oxidative stress-induced cell death. *Mol. Biol. Cell* **2010**, *22*, 128–140. [CrossRef] [PubMed]
24.   Perkins, N.D. Integrating cell-signalling pathways with NF-kappaB and IKK function. *Nat. Rev. Mol. Cell Biol.* **2007**, *8*, 49–62. [CrossRef] [PubMed]
25.   Cai, C.; Yao, Z. Activation of NF-κB in human breast cancer and its role in cell proliferation and progression. *Chin. J. Clin. Oncol.* **2006**, *3*, 5–10. [CrossRef]
26.   Zhao, X.; Sun, X.; Gao, F.; Luo, J.; Sun, Z. Effects of ulinastatin and docataxel on breast tumor growth and expression of IL-6, IL-8, and TNF-α. *J. Exp. Clin. Cancer Res.* **2011**, *30*, 22. [CrossRef] [PubMed]
27.   Park, M.; Hong, J. Roles of NF-κB in cancer and inflammatory diseases and their therapeutic approaches. *Cells* **2016**, *5*, 15. [CrossRef] [PubMed]
28.   Shostak, K.; Chariot, A. NF-κB, stem cells and breast cancer: The links get stronger. *Breast Cancer Res.* **2011**, *13*, 214. [CrossRef] [PubMed]

29. Becker-Weimann, S.; Xiong, G.; Furuta, S.; Han, J.; Kuhn, I.; Akavia, U.-D.; Pe'er, D.; Bissell, M.J.; Xu, R. NFkB disrupts tissue polarity in 3D by preventing integration of microenvironmental signals. *Oncotarget* **2013**, *4*, 2010–2020. [CrossRef] [PubMed]

30. Bissell, M.J.; Radisky, D. Putting tumors in context. *Nat. Rev. Cancer* **2001**, *1*, 46–54. [CrossRef] [PubMed]

31. Mestdagt, M.; Polette, M.; Buttice, G.; Noël, A.; Ueda, A.; Foidart, J. Transactivation of MCP-1/CCL2 by beta-catenin/TCF-4 in human breast cancer cells. *Int. J. Cancer* **2006**, *118*, 35–42. [CrossRef] [PubMed]

32. Keith, B.D. Systematic review of the clinical effect of glucocorticoids on nonhematologic malignancy. *BMC Cancer* **2008**, *8*, 84. [CrossRef] [PubMed]

33. Skor, M.; Wonder, E.; Kocherginsky, M.; Goyal, A.; Hall, B.; Cai, Y.; Conzen, S. Glucocorticoid receptor antagonsims as a novel therapy for triple-negative breast cancer. *Clin. Cancer Res.* **2012**, *100*, 130–134.

34. Mitre-Aguilar, I.B.; Cabrera-Quintero, A.J.; Zentella-Dehesa, A. Genomic and non-genomic effects of glucocorticoids: Implications for breast cancer. *Int. J. Clin. Exp. Pathol.* **2015**, *8*, 1–10. [PubMed]

35. McKay, L.I.; Cidlowski, J.A. Molecular control of immune/inflammatory responses: Interactions between nuclear factor-κB and steroid receptor-signaling pathways. *Endocr. Rev.* **1999**, *20*, 435–459. [CrossRef] [PubMed]

36. Schmidt, S.; Rainer, J.; Ploner, C.; Presul, E.; RimL, S.; Kofler, R. Glucocorticoid-induced apoptosis and glucocorticoid resistance: Molecular mechanisms and clinical relevance. *Cell Death Differ.* **2004**, *11*, 45–55. [CrossRef] [PubMed]

37. Lin, K.T.; Wang, L.H. New dimension of glucocorticoids in cancer treatment. *Steroids* **2016**, *111*, 84–88. [CrossRef] [PubMed]

38. Buxant, F.; Kindt, N.; Laurent, G.; Noel, J.; Saussez, S. Antiproliferative effect of dexamethasone in the MCF-7 breast cancer cell line. *Mol. Med. Rep.* **2015**, 4051–4054. [CrossRef] [PubMed]

39. Mehmeti, M.; Allaoui, R.; Bergenfelz, C.; Saal, L.H.; Ethier, S.P.; Johansson, M.E.; Jirström, K.; Leandersson, K. Expression of functional toll like receptor 4 in estrogen receptor/progesterone receptor-negative breast cancer. *Breast Cancer Res.* **2015**, *17*, 130. [CrossRef] [PubMed]

40. Akira, S.; Uematsu, S.; Takeuchi, O. pathogen recognition and innate immunity. *Cell* **2006**, *3*, 783–801. [CrossRef] [PubMed]

41. Tak, P.P.; Firestein, G.S. NF-κB: A key role in inflammatory diseases. *J. Clin. Investig.* **2001**, *107*, 7–11. [CrossRef] [PubMed]

42. Pahl, H.L. Activators and target genes of Rel/NF-κB transcription factors. *Oncogene* **1999**, *18*, 6853–6866. [CrossRef] [PubMed]

43. Kumar, A.; Takada, Y.; Boriek, A.; Aggarwal, B. Nuclear factor-κB: Its role in health and disease. *J. Mol. Med.* **2004**, *82*, 434–448. [CrossRef] [PubMed]

44. Schütze, S.; Potthoff, K.; Machleidt, T. TNF activates NF-KB by phosphatidylcholine-specific C-induced "Acidic" sphingomyelin breakdown. *Cell* **1992**, *71*, 765–776. [CrossRef]

45. Miyake, K. Innate recognition of lipopolysaccharide by Toll-like receptor 4-MD-2. *Trends Microbiol.* **2004**, *12*, 186–192. [CrossRef] [PubMed]

46. Wajant, H.; Scheurich, P. TNFR1-induced activation of the classical NF-KB pathway. *FEBS J.* **2011**, *278*, 862–876. [CrossRef] [PubMed]

47. Correa, H.; Aristizabal, F.; Duque, C.; Kerr, R. Cytotoxic and antimicrobial activity of pseudopterosins and seco-pseudopterosins isolated from the octocoral *Pseudopterogorgia elisabethae* of san andrés and providencia islands (Southwest Caribbean Sea). *Mar. Drugs* **2011**, *9*, 334–344. [CrossRef] [PubMed]

48. Ata, A.; Win, H.Y.; Holt, D.; Holloway, P.; Segstro, E.P.; Jayatilake, G.S. New antibacterial diterpenes from *Pseudopterogorgia elisabethae*. *Helv. Chim. Acta* **2004**, *87*, 1090–1098. [CrossRef]

49. Yamamoto, M.; Taguchi, Y.; Ito-kureha, T.; Semba, K.; Yamaguchi, N.; Inoue, J. NF-κB non-cell-autonomously regulates cancer stem cell populations in the basal-like breast cancer subtype. *Nat. Commun.* **2013**, *4*, 2299. [CrossRef] [PubMed]

50. Yamaguchi, N.; Ito, T.; Azuma, S.; Ito, E.; Honma, R.; Yanagisawa, Y.; Nishikawa, A.; Kawamura, M.; Imai, J.; Watanabe, S.; et al. Constitutive activation of nuclear factor-kappaB is preferentially involved in the proliferation of basal-like subtype breast cancer cell lines. *Cancer Sci.* **2009**, *100*, 1668–1674. [CrossRef] [PubMed]

51. Ling, J.; Kumar, R. Crosstalk between NFkB and glucocorticoid signaling: A potential target of breast cancer therapy. *Cancer Lett.* **2012**, *322*, 119–126. [CrossRef] [PubMed]

52. Pires, B.R.B.; Mencalha, A.L.; Ferreira, G.M.; de Souza, W.F.; Morgado-Díaz, J.A.; Maia, A.M.; Corrêa, S.; Abdelhay, E.S.F.W. NF-kappaB is involved in the regulation of EMT genes in breast cancer cells. *PLoS ONE* **2017**, *12*, e0169622. [CrossRef] [PubMed]

53. Gordon, A.H.; O'Keefe, R.J.; Schwarz, E.M.; Rosier, R.N.; Puzas, J.E. Nuclear factor-kB-dependent mechanisms in breast cancer cells regulate tumor burden and osteolysis in bone. *Am. Assoc. Cancer Res.* **2005**, *65*, 3209–3218. [CrossRef] [PubMed]

54. Rutz, H.P. Effects of corticosteroid use on treatment of solid tumours. *Lancet* **2002**, *360*, 1969–1970. [CrossRef]

55. Pan, D.; Kocherginsky, M.; Conzen, S.D. Activation of the glucocorticoid receptor is associated with poor prognosis in estrogen receptor-negative breast cancer. *Cancer Res.* **2011**, *71*, 6360–6370. [CrossRef] [PubMed]

56. Mondal, S.K.; Jinka, S.; Pal, K.; Nelli, S.; Dutta, S.K.; Wang, E.; Ahmad, A.; AlKharfy, K.M.; Mukhopadhyay, D.; Banerjee, R. Glucocorticoid receptor-targeted liposomal codelivery of lipophilic drug and Anti-Hsp90 gene: Strategy to induce drug-sensitivity, EMT-reversal, and reduced malignancy in aggressive tumors. *Mol. Pharm.* **2016**, *13*, 2507–2523. [CrossRef] [PubMed]

57. Radisky, D.C.; Bissell, M.J. NF-kappaB links oestrogen receptor signalling and EMT. *Nat. Cell Biol.* **2007**, *9*, 361–363. [CrossRef] [PubMed]

58. Abduljabbar, R.; Negm, O.H.; Lai, C.-F.; Jerjees, D.A.; Al-Kaabi, M.; Hamed, M.R.; Tighe, P.J.; Buluwela, L.; Mukherjee, A.; Green, A.R.; et al. Clinical and biological significance of glucocorticoid receptor (GR) expression in breast cancer. *Breast Cancer Res. Treat.* **2015**, *150*, 335–346. [CrossRef] [PubMed]

59. Chrousos, G.P.; Kino, T. Intracellular glucocorticoid signaling: A formerly simple system turns stochastic. *Sci. Signal.* **2005**, *2005*, 48. [CrossRef] [PubMed]

60. De Bosscher, K.; Vanden Berghe, W.; Haegeman, G. The interplay between the glucocorticoid receptor and nuclear factor-kB or activator protein-1: Molecular mechanisms for gene repression. *Endocr. Rev.* **2003**, *24*, 488–522. [CrossRef] [PubMed]

61. Kino, T.; De Martino, M.U.; Charmandari, E.; Mirani, M.; Chrousos, G.P. Tissue glucocorticoid resistance/hypersensitivity syndromes. *J. Steroid Biochem. Mol. Biol.* **2003**, *85*, 457–467. [CrossRef]

62. Livak, K.J.; Schmittgen, T.D. Analysis of relative gene expression data using real-time quantitative PCR and $2^{-\Delta\Delta C_T}$ method. *Methods* **2001**, *25*, 402–408. [CrossRef] [PubMed]

63. McCloy, R.A.; Rogers, S.; Caldon, C.E.; Lorca, T.; Castro, A.; Burgess, A. Partial inhibition of Cdk1 in G2 phase overrides the SAC and decouples mitotic events. *Cell Cycle* **2014**, *13*, 1400–1412. [CrossRef] [PubMed]

marine drugs

MDPI

Article

Selenium-Containing Polysaccharide-Protein Complex in Se-Enriched *Ulva fasciata* Induces Mitochondria-Mediated Apoptosis in A549 Human Lung Cancer Cells

Xian Sun [1,2,†], **Yu Zhong** [1,2,†], **Hongtian Luo** [1,2] and **Yufeng Yang** [1,2,*]

1   Institute of Hydrobiology, Jinan University, Jinan 510632, China; imytian@163.com (X.S.); zhongyu@pku.edu.cn (Y.Z.); lhtcoffee@163.com (H.L.)

2   Key Laboratory of Aquatic Eutrophication and Control of Harmful Algal Blooms, Guangdong Higher Education Institutes, Guangzhou 510632, China

*   Correspondence: tyyf@jnu.edu.cn.; Tel./Fax: +86-020-8522-1397

†   These authors contributed equally to this work.

Received: 23 March 2017; Accepted: 1 July 2017; Published: 16 July 2017

Abstract: The role of selenium (Se) and *Ulva fasciata* as potent cancer chemopreventive and chemotherapeutic agents has been supported by epidemiological, preclinical, and clinical studies. In this study, Se-containing polysaccharide-protein complex (Se-PPC), a novel organoselenium compound, a Se-containing polysaccharide-protein complex in Se-enriched *Ulva fasciata*, is a potent anti-proliferative agent against human lung cancer A549 cells. Se-PPC markedly inhibited the growth of cancer cells via induction of apoptosis which was accompanied by the formation of apoptotic bodies, an increase in the population of apoptotic sub-G1 phase cells, upregulation of p53, and activation of caspase-3 in A549 cells. Further investigation on intracellular mechanisms indicated that cytochrome C was released from mitochondria into cytosol in A549 cells after Se-PPC treatment. Se-PPC induced depletion of mitochondrial membrane potential ($\Delta\Psi m$) in A549 cells through regulating the expression of anti-apoptotic (Bcl-2, Bcl-XL) and pro-apoptotic (Bax, Bid) proteins, resulting in disruption of the activation of caspase-9. This is the first report to demonstrate the cytotoxic effect of Se-PPC on human cancer cells and to provide a possible mechanism for this activity. Thus, Se-PPC is a promising novel organoselenium compound with potential to treat human cancers.

Keywords: *Ulva fasciata*; selenium-containing polysaccharide-protein complex; apoptosis; mitochondria; reactive oxygen species

---

1. Introduction

Due to the increasing incidence of cancer in both developing and developed countries, new chemotherapy compounds are needed [1]. Employing natural or synthetic agents to prevent or suppress the progression of invasive cancers has recently been recognized as an approach with enormous potential [2]. Seaweeds (marine algae) are rich in dietary fiber, minerals, lipids, proteins, omega-3 fatty acids, essential amino acids, polysaccharides, and vitamins A, B, C, and E [3–6] Studies on the bioactivities of seaweeds reveal numerous health-promoting effects, including anti-oxidative, anti-inflammatory, antimicrobial, and anti-cancer effects. These studies have indicated that marine algae constitute a promising source of novel compounds with potential as human therapeutic agents.

Recently, polysaccharides (PS) from marine organisms have garnered attention because of their potential to be used as ingredients in new medicines and food [7]. PS, including the polysaccharide-protein complex (PPC), are major bioactive constituents of seaweeds with a range of anti-tumor, immune-modulatory, and antioxidant effects [8,9]. However, among marine macrophytes, marine green algae have been less studied than brown and red algae as sources of PPC with such effects. Their antitumor properties have, however, been reported, mainly for those of ulvans. Tabarsa et al. [6] reported that ulvans from *Ulva pertusa* showed little cytotoxicity against tumor cells, but significantly stimulated immunity by inducing nitric oxide and cytokine production.

Selenium (Se) plays an important role in many physiological processes and is therefore an essential trace element for human beings and animals [10] However, organic Se is absorbed more readily and is less toxic than inorganic Se. Se-polysaccharide is reported to be more potent than either Se or polysaccharide. For example, selenylated polysaccharides show greater antioxidant activity than native polysaccharides [11]. Natural plant polysaccharides generally have a low content of Se even in plants grown in high-selenium soil and do not provide adequate dietary Se [12]. Therefore, the use of bioenrichment to prepare high Se polysaccharide is well established and applied by many researchers [13,14]. In our previous study, we found that if *Ulva fasciata* is grown in 500 mg Se/L it can be a source of Se-enriched food because more than 80% of inorganic Se was transformed into the organic form [14].

There is accumulating evidence that bioactive compounds from algae have anticancer effects by inhibition of cancer cell growth, as well as invasion and metastasis. They also induce apoptosis in cancer cells [8]. Apoptosis, programmed cell death, can be induced by both the death receptor and mitochondrial pathways [15]. Apoptotic signals are mediated by Bcl-2 family members, including the anti-apoptotic proteins Bcl-2 and Bcl-xL, and the pro-apoptotic proteins—Bax, Bak, and Bad—in the mitochondrial pathway [16]. The key process of mitochondria-mediated apoptosis is the collapse of mitochondrial membrane potential, which is followed by the translocation of cytochrome c from the mitochondria into the cytosol [17]. Then the subsequent activation of caspases was allowed [18]. The caspase-9 and caspase-3 activated forms are among the main mediators of apoptosis. The two enzymes cleave a wide range of important proteins, including other caspases and the anti-apoptotic protein (such as Bcl-2) [19].

Ulva fasciata, also known as sea lettuce, species of the green algal genus *Ulva*, grows abundantly along coastal seashores. Despite the evidence for some biological effect of *Ulva fasciata* against colon cancer cells, there are no available reports of an antitumor effect of Se-PPC from Se-enriched *Ulva fasciata*. Thus, in the present in vitro study, the cytotoxic effect of this Se-PPC on A549 human lung cancer cells was investigated. We aimed to uncover the cytotoxic mechanism of reactive oxygen species (ROS) and mitochondrial apoptosis using various molecular and cellular techniques.

2. Materials and Methods

2.1. Materials

Sodium selenite (Na_2SeO_3), 3-(4,5-dimethylthiazol-2-yl)-2,5-diphenyltetrazolium bromide (MTT), propidium iodide (PI), bicinchoninic acid (BCA) for the protein determination kit, and 2′,7′-dichlorofluorescein diacetate (DCF-DA) were purchased from Sigma (St. Louis, MO, USA). Caspase-3 substrate (Ac-DEVDAMC) was purchased from Biomol (Hamburg, Germany). Caspase-9 substrate (Ac-LEHD-AFC) and caspase-8 substrate (IETD-AFC) were purchased from Calbiochem (San Diego, CA, USA). The primary antibodies used against Cyclin D1, CDK4 p53, Fas, Bax, Bid, Bcl-2, Bcl-XL, and β-actin, were purchased from Santa Cruz Biotechnology (Santa Cruz, TX, USA). The ultrapure water used in all experiments, supplied by a Milli-Q water purification system from Millipore (Billerica, MA, USA).

For the other assays, cells were seeded in 12-well plates at a density of 6×10^5 cells/well.

2.2. Preparation, Extraction, and Isolation of Se-PPC

Ulva fasciata was collected from the Nanao Island Cultivation Zone (116.6° E, 23.3° N), Shantou, Guangdong, China. Before Se-enriched treatments, the seaweed was acclimated in sterilized seawater for four weeks. Throughout the study, the *Ulva fasciata* was maintained in sterilized seawater enriched by 100 μM of NaNO$_3$-N and 10 μM of NaH$_2$PO$_4$-P at 20 ± 0.5 °C under cool-white fluorescent lamps at 80 μmol photons m^2 s^{-1}. All solutions and glassware were autoclaved at 121 °C for 15 min prior to use.

The seaweed was cultured in 2 L Erlenmeyer flasks containing 1.5 L medium supplemented with Na$_2$CO$_3$ and sterile air containing 2% CO$_2$ as the carbon sources. Se was added in the form of sodium selenite (Na$_2$SeO$_3$) at concentrations of 500 mg/L. 5 g *Ulva fasciata* FW samples of were placed in each flask which was covered by gauze and placed in indoor tanks at 20 °C, under a light intensity of 275 μmol photons m^2 s^{-1}, at pH 8.0, 30 PSU salinity, and with 12 h:12 h light-dark cycle. Before Se-PPC extraction and isolation, the *Ulva fasciata* thalli were washed three times carefully with ultrapure water to remove the surface Se.

Ultrasound-assisted extraction (UAE) was performed with a Model VCX-130 ultrasonic processor with a probe horn of 20 kHz frequency and 130 W power (Sonics & Materials Inc., Newton, MA, USA). A 12 mm diameter horn tip was used in the UAE experiments with the power fixed at 70% amplitude (corresponding to an intensity of 26.5 W/cm^2 tip surface) and the total irradiation period at 60 min (to achieve the maximum Se-PPC yield according to preliminary tests). Each 3 g *Ulva fasciata* sample was mixed with 90 mL of distilled water in a 250 mL plastic centrifuge bottle; the ultrasonic probe was inserted into the sample liquid at ca. 2 cm depth. The sample bottle was immersed in ice with the maximum temperature below 50 °C throughout the UAE period.

The liquid extract was separated from the solid residues by centrifugation (6000 rpm, 10 min) and subjected to ethanol precipitation (80%, v/v ethanol) as reported previously [20]. The precipitates were collected after 16,000 rpm, 15 min centrifugation and lyophilized, giving the (crude) Se-PPC fraction. The Se content in Se-PPC was determined by ICP-AES following Sun et al. (2014).

2.3. Cell Lines and Cell Culture

In this study, A549 human lung cancer cells and HK-2 human renal tubular epithelial cells were obtained from American Type Culture Collection (ATCC, Rockville, MD, USA). All cells were cultured in 75 cm^2 culture flasks in RPMI 1640 (Roswell Park Memorial Institute 1640, Invitrogen, Carlsbad, CA, USA) (for A549 and HK-2) culture medium supplemented with 10% fetal bovine serum (Hyclone, Waltham, MA, USA), 100 units/mL penicillin and 50 units/mL streptomycin in a humidified incubator with an atmosphere of 95% air and 5% CO$_2$ at 37 °C. After growth to confluence, the cells were detached with a 0.25% trypsin for passage, and the cells were ready for study until the cell growth was in a stable state and in the logarithmic growth phase unless otherwise specified.

2.4. Cell Viability Examination

The effect of Se-PPC on cell proliferation was determined by the MTT assay. Cells were seeded in 96-well tissue culture plates at 3.0 × 10^3 cells/well for 24 h. The cells were then incubated with Se-PPC at different concentrations for 72 h. After incubation, 20 μL of MTT solution (5 mg/mL phosphate buffered saline) was added to each well and incubated for 5 h. The medium was aspirated and replaced with 150 mL/well of acidic isopropanol (0.04 N HCl in isopropanol) to dissolve the formazan salt formed. The color intensity of the formazan solution, which reflects the cell growth condition, was measured at 570 nm using a microplate spectrophotometer (SpectroAmaxTM 250, VARIAN, Palo Alto, CA, USA).

2.5. Flow Cytometric Analysis

Cell cycle distribution was monitored by flow cytometry. Briefly, cells treated with or without Se-PPC were harvested by centrifugation and washed with PBS. Cells were stained with PI after fixation with 70% ethanol at −20 °C overnight. Labelled cells were washed with PBS and then analyzed by the flow cytometer. The cell cycle distribution was analyzed using MultiCycle software (Phoenix Flow Systems, San Diego, CA, USA). The proportions of cells in G0/G1, S, and G2/M phases were represented in DNA histograms. Apoptotic cells with hypodiploid DNA content were measured by quantifying the sub-G1 peak. For each experiment, 10,000 events per sample were recorded.

2.6. Caspase Activity Assay

Harvested cell pellets were suspended in cell lysis buffer and incubated on ice for 1 h. After centrifugation at $11,000 \times g$ for 30 min, supernatants were collected and immediately measured for protein concentration and caspase activity. Briefly, cell lysates were placed in 96-well plates and then specific caspase substrates (Ac-DEVD-AMC for caspase-3, Ac-IETD-AMC for caspase-8, and Ac-LEHD-AMC for caspase-9) were added. Plates were incubated at 37 °C for 1 h. Caspase activity was determined by fluorescence intensity under the excitation and emission wavelengths set at 380 and 440 nm, respectively.

2.7. Evaluation of Mitochondrial Membrane Potential ($\Delta\Psi m$)

Cells in 6-well plates were trypsinized and resuspended in 0.5 mL of PBS buffer containing 10 µg/mL of JC-1. After incubation for 10 min at 37 °C in the incubator, cells were immediately centrifuged to remove the supernatant. Cell pellets were suspended in PBS and then analyzed by flow cytometry. The percentage of the green fluorescence from JC-1 monomers was used to represent the cells that lost $\Delta\Psi m$.

2.8. Western Blot Analysis

First RIPA lysis buffer (50 mM TriseHCl, 150 mM NaCl, 0.1% SDS, 1% NP-40, 0.5% sodium deoxycholate, 1 mM PMSF, 100 mM leupeptin, and 2 mg/mL aprotinin, pH 8.0) was used to extract total cellular proteins and then the protein extracts were resolved by loading equal amounts of protein, in 10% SDS-PAGEE gel, per lane. They were then put onto Immobilon-P PVDF transfer membranes (Millipore, Bedford, MA, USA) by electroblotting. As a final step, they were blocked with 5% non-fat milk in TBST on a shaker at RT for 1 h.

After that, the membranes were probed by primary antibodies (Cell signaling, Danvers, MA, USA) diluted 1:1000 in 5% nonfat milk at 4 °C overnight, and by secondary antibodies, conjugated with horseradish peroxidase at 1:2000 dilution, at RT for 1 h. To assess the presence of comparable amounts of protein in each lane, the membranes were stripped to detect b-actin (Proteintech group, Chicago, IL, USA). All the protein bands were developed by the SuperSignal West Pico kit (Pierce Biotechnology, Rockford, IL, USA).

2.9. Assay for Mitochondrial Cytochrome C Release

This assay was performed according to cytochrome C releasing apoptosis assay kit's (Biovision, San Francisco, CA, USA) instructions. In brief, after treatment, 1×10^6 cells were pelleted by centrifugation and washed twice with ice-cold PBS. Cell pellets were resuspended with 1 mL cytosol extraction buffer mix containing DTT and protease inhibitor, and incubated for 10 min on ice. After homogenization, unbroken cells and large debris were removed by centrifugation. The resulting supernatants were saved as cytosolic extracts at −70 °C. The pellets were resuspended with 100 mL extraction buffer mix containing DTT and protease inhibitor, and saved as mitochondrial fractions. We loaded 30 mg cytosolic and mitochondrial fractions isolated from A549 cells on 12% SDS-PAGE. Then western blot proceeded with cytochrome C antibody (Biovision, San Francisco, CA, USA).

2.10. Measurement of ROS Generation

The effects of Se-PPC on ROS-initiated intracellular oxidation were evaluated by the DCF fluorescence assay. Briefly, cells were harvested, washed with PBS, and suspended in PBS (1×10^6 cells/mL) containing 10 mM of DCFH-DA. After incubation at 37 °C for 30 min, cells were collected and resuspended in PBS. Then, the ROS level was determined by measuring the fluorescence intensity on a Tecan SAFIRE multifunctional monochromator-based microplate reader, with excitation and emission wavelengths of 500 and 529 nm, respectively. Experiments were performed in triplicate.

2.11. Statistical Analysis

Experiments were carried out at least in triplicate and results were expressed as mean ± SD. Statistical analysis was performed using the SPSS statistical package (SPSS 13.0 for Windows; SPSS, Inc., Chicago, IL, USA). The difference between two groups was analyzed by the two-tailed Student's *t*-test, and between three or more groups by one-way ANOVA multiple comparisons. A difference with $p < 0.05$ (*) or $p < 0.01$ (**) was considered statistically significant.

3. Results and Discussion

3.1. Cytotoxic Effects of Se-PPC on Various Human Cancer and Normal Cell Lines

Many organic selenocompounds have been reported to have potent chemopreventive activities [21]. However, the balance between the therapeutic potential and the toxic effect of a compound is very important when evaluating its pharmacological usefulness. In this study, in vitro cytotoxicities of Se-PPC to A549 cells and normal cells were compared. Se-PPC, from the Se-enriched green seaweed, *Ulva fasciata*, contained 44.4 µg/g Se. The anti-proliferative activities of Se-PPC were first screened against human lung cancer A549 cells in a dose-dependent manner by MTT assay (Figure 1). After the 72 h treatment with Se-PPC at doses of 3, 4, 5 and 6 µg/mL, the percentage of surviving A549 cells were significantly reduced to 37.99, 25.09, 15.26, and 13.05% of the control, respectively (Figure 1A). Exposure for 72 h to 3 µg Se-PPC /mL induced 62.01% A549 (Figure 1A). Se-PPC exhibited broad inhibition on A549 cancer cells with the IC_{50} values of 2.8 µg/mL (Figure 1B). Despite this potency, Se-PPC showed low cytotoxicity toward normal cells (HK-2 renal tubular epithelial cells) with an IC_{50} value of 27.7 µg/mL. These results suggest that Se-PPC selects between cancer and normal cells and has, therefore, potential application in cancer chemoprevention and chemotherapy.

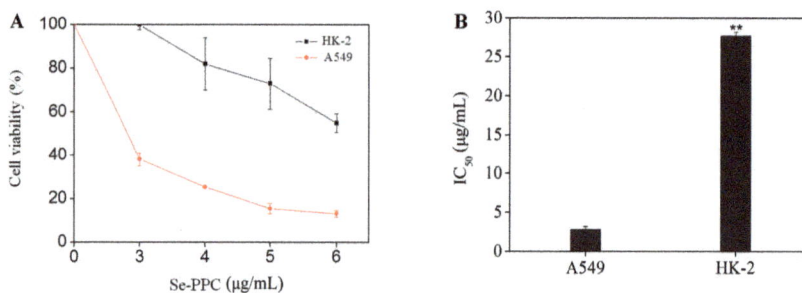

Figure 1. (A) Cytotoxic effects of Selenium-containing polysaccharide-protein complex (Se-PPC) on A549 human lung cancer cells and normal cells (HK-2 human renal tubular epithelial cells). Data are expressed as the decrease in cell viability (for MTT assay); **(B)** Growth inhibition of Se-PPC was expressed as the IC_{50} (for MTT assay). Each IC_{50} value represents the mean ± SD of three independent experiments. Cells were treated with Se-PPC for 72 h. All values were obtained at least from three independent experiments. Difference between normal cells and cancer cells with $p < 0.01$ (**) was considered statistically significant.

3.2. Apoptosis-Inducing Activities of Se-PPC and the Underlying Mechanisms

The inhibition of proliferation of cells treated by Se-PPC could be either the induction of apoptosis or cell cycle arrest, or a combination of the two. The role of apoptosis in the action of anticancer drugs has become clearer [22]. Though, we investigated the underlying mechanism of Se-PPC-induced cell death. El-Bayoumy & Sinha. [21] reported that apoptosis could be critical for cancer chemoprevention by selenocompounds. Our flow cytometry revealed that exposure of the A549 cells to different concentrations of Se-PPC results in a dose-dependent increase in the proportion of apoptotic cells as reflected by the Sub-G1 populations (7.3–61.3%) with a treatment of 2–16 μg/mL Se-PPC (Figure 2A). Moreover, no significant changes in G0/G1, S, and G2/M phases were observed in treated cells. To investigate the potential mechanisms of Se-PPC-mediated induction of cell cycle arrest, the effects of Se-PPC on the expression of CDK4 and Cyclin D1, which are necessary for cell cycle progression, were evaluated. A549 cells were treated with various concentrations of Se-PPC (2–16 μg/mL), and the expression levels of Cyclin D1 and CDK4 proteins were analyzed by Western blot analysis. Se-PPC significantly decreases the protein levels of CDK4 and Cyclin D1 in A549 cells (Figure 3B). In addition, relative to untreated control, Se-PPC suppressed the levels of Cyclin D1 and CDK4 in ovarian cancer cells in a dose-dependent manner ($p < 0.05$). In mammalian cells, cell cycle progression is tightly regulated through the activation of CDKs whose association with the corresponding regulatory cyclins is required for their activation [23]. It is well known that G1 to S phase transition is regulated by complexes formed by Cyclin D and CDK4 [24]. These results indicate that cell death induced by the Se-PPC is mainly due to the induction of apoptosis caused by cell cycle arrest.

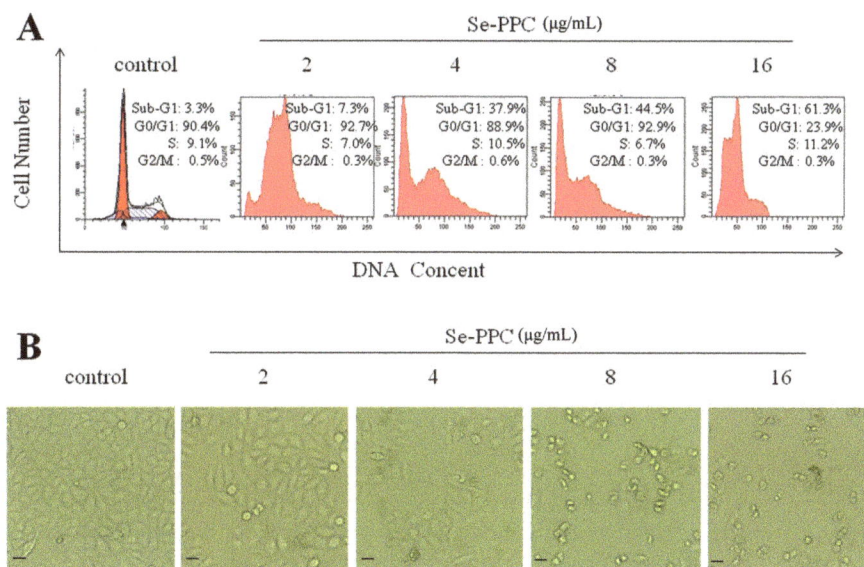

Figure 2. (**A**) Effects of Se-PPC on cell apoptosis and cell cycle distribution in A549 cells (scale bar: 50 μm). The cells treated with different concentrations of Se-PPC for 72 h were collected and stained with PI after fixation, and then analyzed by flow cytometry. Cellular DNA histograms were analyzed by the MultiCycle software. Each value represents the mean of three independent experiments; (**B**) Morphological changes of A549 cells treated with Se-PPC for 72 h observed by phase-contrast microscopy (magnification, 100×). The images shown here are representative of three independent experiments with similar results.

Phase-contrast observations showed that A549 cells treated with Se-PPC exhibited a dose-dependent reduction in cell numbers, a loss of cell-to-cell contact, cell shrinkage, and formation of apoptotic bodies (Figure 2B). Also, the density of cells decreased with the 2–16 μg/mL Se-PPC treatment. Furthermore, when the cancer cells were treated with high concentrations of Se-PPC

(16 µg/mL), most of the cells coalesced and were suspended in the culture medium. Se-PPC induced a change in cell morphology and inhibited cancer cell growth in a dose-dependent manner.

The release of cytochrome c from the mitochondria to the cytosol is one of the early events that subsequently lead to apoptosis by activation of caspases, including caspase-3 [25]. The release of cytochrome c into cytosol leads to activation of procaspase-9 in the apoptosome and then causes the cleavage of caspase-3 [26]. Caspase-3, believed to be an important effector protease, is cleaved and activated during apoptosis [25]. For these reasons, we examined the effector caspase (caspase-3) by spectrophotometry; western blot analysis was performed to detect the effect of Se-PPC on the p53 in A549 cells. Our results showed that caspase-3 activities (2.42–3.48 folds) increased significantly ($p < 0.05$) compared to the control, and Se-PPC upregulated the expression of p53 (Figure 3). In cell models, DNA damage activates ATM (ataxia telangiectasia mutated) and ATR (ataxia telangiectasia and Rad3 related) proteins, which signal downstream to checkpoint kinases, such as CHK1 and CHK2. Also, the tumor suppressor gene, p53, which is a major player in the apoptosis because it induces transcription of pro-apoptotic factors and inhibition of pro-survival factors [27]. These results suggest that caspase-3 and p53 contribute to cell apoptosis induced by Se-PPC.

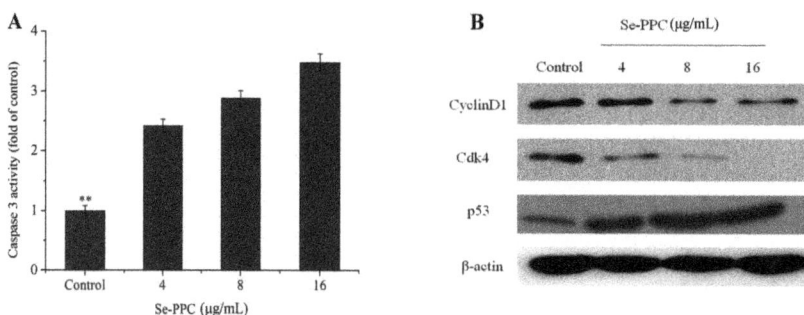

Figure 3. (**A**) Effect of Se-PPC on caspase-3 activity of A549 cells; (**B**) Effect of Se-PPC on cyclin-dependent kinase 4, Cyclin D1, and p53 protein expression of A549 cells. The values represent means ± SD of triplicate determinations. Difference between treatment and control cells with $p < 0.01$ (**) was considered statistically significant.

3.3. Mitochondria Plays an Important Role in Se-PPC-Induced Apoptosis

Generally, apoptosis occurs via death receptor-dependent (extrinsic) or mitochondrial (intrinsic) pathways. The mitochondrial pathway of cell death is mediated by Bcl-2 family proteins which disrupt the mitochondria membrane potential and result in release of apoptogenic factors such as cytochrome c, Smac/Diablo, and AIF, into the cytosol [28]. Cytochrome c then forms an apoptosome containing apoptosis activating factor 1 and caspase-9, which then activates the downstream apoptotic signals [22]. In this content, we examined the cytochrome c levels in cytosol fractions of cells treated with Se-PPC at doses of 4, 8, and 16 µg/mL. The result of western blot showed that the cytosolic cytochrome c protein expression increased markedly in a dose-dependent manner in Se-PPC-treated cells (Figure 4B). Since release of cytochrome c into the cytosol is usually preceded or accompanied by a loss or disruption of mitochondrial membrane potential and this collapse is an essential step occurring in cells undergoing apoptosis [29]. Loss of $\Delta\Psi m$ is associated with the activation of caspases and the initiation of apoptotic cascades. Thus, the status of mitochondria in A549 cells exposed to Se-PPC was investigated by JC-1 flow cytometric analysis. It was found that Se-PPC induced significant depletion of $\Delta\Psi m$ in A549 cells (Figure 4C). The percentage of cells with depolarized mitochondria increased from 6.80% (control) to 16.55% (4 µg/mL), 30.40% (8 µg/mL), and 46.09% (16 µg/mL), respectively.

Caspases are a family of cysteine proteases that play central roles in the initiation and execution of apoptosis [21] The extrinsic pathway is triggered by activation of death receptors. The formation of a death-inducing signaling complex subsequently activates initiator caspase-8 [30]. Because caspases have been identified as targets for therapeutic intervention using fluorimetry, we measured the activity of two initiator caspases, caspase-8 (Fas/TNF-mediated) and caspase-9 (mitochondrial-mediated). Our results showed that Se-PPC-evoked apoptosis resulted in dose-dependent activation of caspase-8 and caspase-9 in A549 cells, suggesting that both caspase-8 and caspase-9 were involved in Se-PPC-induced apoptosis (Figure 4A). Activity of initiator caspase-9 increased 1.75–2.74-fold in cells exposed to 4–16 mM of Se-PPC compared with controls. In contrast, little increase in activities of caspases-8 (1.08–1.66-fold) in response to Se-PPC treatment was observed. Meanwhile, under Se-PPC treatments, the expression of Fas in A549 cells were upregulated insignificantly (Figure 4B), suggesting that the contribution of caspase-8 to the induction of cell apoptosis was likely to be insignificant, and that the mitochondrial-mediated apoptotic pathway played the major role in Se-PPC-induced apoptosis in A549 cells. Because it was well-known that caspase-8 is activated via the death receptor-mediated pathways.

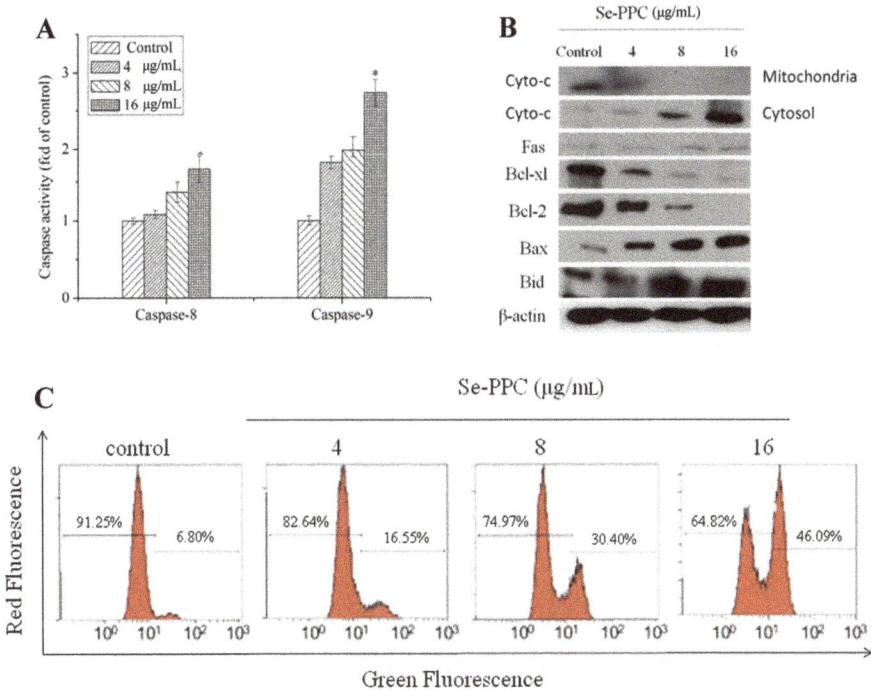

Figure 4. (**A**) Effect of Se-PPC on caspase-8 and caspase-9 activities of A549 cells; (**B**) Effect of Se-PPC on cytochrome C, Fas, Bcl-2, Bcl-XL, Bid, and Bax protein expression of A549 cells; (**C**) Cells treated with Se-PPC were harvested and stained with the mitochondria-selective dye JC-1 and then analyzed by flow cytometry. The number in the right region of each dot plot represents the percentage of cells that emit green fluorescence due to the depletion of $\Delta\Psi m$. The values represent means $\pm$ SD of triplicate determinations. Difference between treatment and control cells with $p < 0.05$ (*) was considered statistically significant.

The Bcl-2 family was divided into two major categories, namely anti-apoptotic proteins (Bcl-2 and Bcl-XL) and pro-apoptotic proteins (Bax and Bid) [31]. Pro-survival family members associate with the mitochondrial outer membrane and maintain their integrity. In contrast, pro-apoptotic members such as Bax and Bid oligomerize in the outer membranes of the mitochondria and disrupt their integrity, causing the release of apoptogenic factors [28]. In this study, under Se-PPC treatments, the expression of Bax and Bid in A549 cells were upregulated, but the expression of Bcl-2 and Bcl-XL were downregulated (Figure 4B). These results indicated that Se-PPC induces mitochondria-mediated apoptosis by regulating the expression of Bcl-2 family proteins.

3.4. Oxidative Stress Is Involved in Se-PPC-Induced Apoptosis

Many chemopreventive and chemotherapeutic agents have been found to induce cancer cell apoptosis through upregulation of intracellular reactive oxygen species (ROS) generation [32]. Letavayova, Vlckova & Brozmanova [33] suggested that the toxicity of Se is due to the induction of oxidative stress and disruption of redox homeostasis. The mitochondrial respiratory chain is a potential source of ROS [34]. ROS—including the superoxide anion, hydrogen peroxide, and hydroxyl radical—are produced under normal aerobic growth conditions within cells, but they are elevated under the influence of external stimuli. Intracellular ROS may attack cellular membrane lipids, proteins, and DNA and also cause oxidative injury [35]. Considerable evidence has suggested that DNA damage can cause cell death by induction of apoptosis via various signaling pathways [27]. Some selenocompounds have been reported to have the potential to induce DNA damage [33]. Nilsonne et al. [36] have provided evidence that ROS generation acts as an important cellular event induced by Se compounds and results in cell apoptosis and/or cell cycle arrest. The involvement of oxidative stress in Se-PPC-induced apoptotic cancer cells was investigated to gain insight into the mechanism of the cytotoxic action of Se-PPC. Our results showed that treatments with Se-PPC generated a dose-dependent increase in DCF fluorescence intensity, indicating upregulation of intracellular ROS levels, suggesting that ROS is a critical mediator in Se-PPC-induced cell apoptosis in A549 cells (Figure 5).

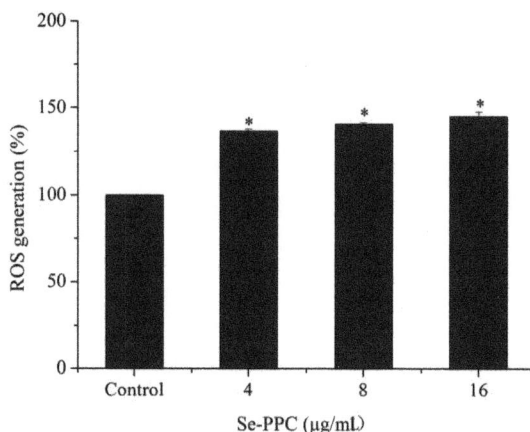

Figure 5. ROS overproduction in A549 cells induced by Se-PPC as determined by DCF fluorescence assay. Cells were treated with indicated concentrations of Se-PPC for 24 h. All experiments were carried out at least in triplicate. Difference between treatments and control with $p < 0.05$ (*) was considered statistically significant.

4. Conclusions

In conclusion, it is shown for the first time that Se-PPC is a novel anti-proliferative agent with a broad spectrum of inhibitions against A549 cancer cells via the induction of apoptosis. However, Se-PPC was found to show a low cytotoxicity toward HK-2 renal tubular epithelial cells. Overproduction of ROS contributes to Se-PPC-induced apoptosis, which in turn leads to apoptotic signals including mitochondria- and caspase-dependent processes in human lung cancer A549 cells. These findings indicate that Se-PPC is a promising organoselenium agent for the treatment of human cancers.

Acknowledgments: This research was supported by the Nature Science Foundation of China (U1301235) and Chinese Special Fund for Agro-scientific Research in the Public Interest (201403008). We are grateful of Larry Liddle (Long Island University, New York, NY, USA) and Zhili He (University of Oklahoma, Oklahoma, OK, USA) for their very valuable comments on the manuscript.

Author Contributions: Yufeng Yang and Yu Zhong conceived and designed the experiments; Yu Zhong and Xian Sun implemented the experiments, and collected and analyzed all of the samples; Yu Zhong, Hongtian Luo and Xian Sun analyzed the data and produced the figures; Xian Sun wrote the paper; Yufeng Yang reviewed the manuscript.

Conflicts of Interest: The authors declare no conflict of interest.

References

1. Pereira, D.M.; Cheel, J.; Areche, C.; San-Martin, A.; Rovirosa, J.; Silva, L.R.; Valentao, P.; Andrade, P.B. Anti-Proliferative Activity of Meroditerpenoids Isolated from the Brown Alga Stypopodium flabelliforme against Several Cancer Cell Lines. *Marinedrugs* **2011**, *9*, 852–862. [CrossRef] [PubMed]
2. Mann, J.R.; Backlund, M.G.; DuBois, R.N. Mechanisms of disease: Inflammatory mediators and cancer prevention. *Nat. Clin. Pract. Oncol.* **2005**, *2*, 202–210. [CrossRef] [PubMed]
3. Cerna, M. Seaweed Proteins and Amino Acids as Nutraceuticals. *Adv. Food Nutr. Res.* **2011**, *64*, 297–312. [PubMed]
4. Misurcova, L.; Skrovankova, S.; Samek, D.; Ambrozova, J.; Machu, L. Health Benefits of Algal Polysaccharides in Human Nutrition. *Adv. Food Nutr. Res.* **2012**, *66*, 75–145. [PubMed]
5. Rajapakse, N.; Kim, S. Nutritional and Digestive Health Benefits of Seaweed. *Adv. Food Nutr. Res.* **2011**, *64*, 17–28. [PubMed]
6. Tabarsa, M.; Rezaei, M.; Ramezanpour, Z.; Waaland, J.R. Chemical compositions of the marine algae Gracilaria salicornia (Rhodophyta) and Ulva lactuca (Chlorophyta) as a potential food source. *J. Sci. Food Agric.* **2012**, *92*, 2500–2506. [CrossRef] [PubMed]
7. Fedorov, S.N.; Ermakova, S.P.; Zvyagintseva, T.N.; Stonik, V.A. Anticancer and Cancer Preventive Properties of Marine Polysaccharides: Some Results and Prospects. *Marinedrugs* **2013**, *11*, 4876–4901. [CrossRef] [PubMed]
8. Farooqi, A.A.; Butt, G.; Razzaq, Z. Algae extracts and methyl jasmonate anti-cancer activities in prostate cancer: Choreographers of 'the dance macabre'. *Cancer Cell Int.* **2012**, *12*. [CrossRef]
9. Zhang, M.; Cui, S.W.; Cheung, P.C.K.; Wang, Q. Polysaccharides from mushrooms: A review on their isolation process, structural characteristics and antitumor activity. *Trends Food Sci. Technol.* **2006**, *18*, 4–19. [CrossRef]
10. Sun, X.; Zhong, Y.; Huang, Z.; Yang, Y. Selenium Accumulation in Unicellular Green Alga Chlorella vulgaris and Its Effects on Antioxidant Enzymes and Content of Photosynthetic Pigments. *PLoS ONE* **2014**, *9*, e11227011. [CrossRef] [PubMed]
11. Guo, Y.; Pan, D.; Li, H.; Sun, Y.; Zeng, X.; Yan, B. Antioxidant and immunomodulatory activity of selenium exopolysaccharide produced by Lactococcus lactis subsp lactis. *Food Chem.* **2013**, *138*, 84–89. [CrossRef] [PubMed]
12. Hou, R.; Chen, J.; Yue, C.; Li, X.; Liu, J.; Gao, Z.; Liu, C.; Lu, Y.; Wang, D.; Li, H.; et al. Modification of lily polysaccharide by selenylation and the immune-enhancing activity. *Carbohydr. Polym.* **2016**, *142*, 73–81. [CrossRef] [PubMed]
13. Zhao, B.; Zhang, J.; Yao, J.; Song, S.; Yin, Z.; Gao, Q. Selenylation modification can enhance antioxidant activity of Potentilla anserina L. polysaccharide. *Int. J. Biol. Macromol.* **2013**, *58*, 320–328. [CrossRef] [PubMed]
14. Zhong, Y.; Chen, T.; Zheng, W.; Yang, Y. Selenium enhances antioxidant activity and photosynthesis in Ulva fasciata. *J. Appl. Phycol.* **2015**, *27*, 555–562. [CrossRef]

15. Li, J.; Huang, C.Y.; Zheng, R.L.; Cui, K.R.; Li, J.F. Hydrogen peroxide induces apoptosis in human hepatoma cells and alters cell redox status. *Cell Biol. Int.* **2000**, *24*, 9–23. [CrossRef] [PubMed]

16. Reuter, S.; Eifes, S.; Dicato, M.; Aggarwal, B.B.; Diederich, M. Modulation of anti-apoptotic and survival pathways by curcumin as a strategy to induce apoptosis in cancer cells. *Biochem. Pharmacol.* **2008**, *76*, 1340–1351. [CrossRef] [PubMed]

17. Balaban, R.S.; Nemoto, S.; Finkel, T. Mitochondria, oxidants, and aging. *Cell* **2005**, *120*, 483–495. [CrossRef] [PubMed]

18. Zamzami, N.; Metivier, D.; Kroemer, G. Quantitation of mitochondrial transmembrane potential in cells and in isolated mitochondria. *Apoptosis* **2000**, *322*, 208–213.

19. Nicholson, D.W. Caspase structure, proteolytic substrates, and function during apoptotic cell death. *Cell Death Differ.* **1999**, *6*, 1028–1042. [CrossRef] [PubMed]

20. Leung, P.H.; Zhao, S.; Ho, K.P.; Wu, J.Y. Chemical properties and antioxidant activity of exopolysaccharides from mycelial culture of Cordyceps sinensis fungus Cs-HK1. *Food Chem.* **2009**, *114*, 1251–1256. [CrossRef]

21. El-Bayoumy, K.; Sinha, R. Mechanisms of mammary cancer chemoprevention by organoselenium compounds. *Mutat. Res.* **2004**, *551*, 181–197. [CrossRef] [PubMed]

22. Kim, R. Recent advances in understanding the cell death pathways activated by anticancer therapy. *Cancer* **2005**, *103*, 1551–1560. [CrossRef] [PubMed]

23. Deshpande, A.; Sicinski, P.; Hinds, P.W. Cyclins and cdks in development and cancer: A perspective. *Oncogene* **2005**, *24*, 2909–2915. [CrossRef] [PubMed]

24. Malumbres, M.; Barbacid, M. Cell cycle, CDKs and cancer: A changing paradigm. *Nat. Rev. Cancer* **2009**, *9*, 153–166. [CrossRef] [PubMed]

25. Riedl, S.J.; Shi, Y.G. Molecular mechanisms of caspase regulation during apoptosis. *Nat. Rev. Mol. Cell Biol.* **2004**, *5*, 897–907. [CrossRef] [PubMed]

26. Oliver, F.J.; de la Rubia, G.; Rolli, V.; Ruiz-Ruiz, M.C.; de Murcia, G.; Murcia, J.M. Importance of poly(ADP-ribose) polymerase and its cleavage in apoptosis. Lesson from an uncleavable mutant. *J. Biol. Chem.* **1998**, *273*, 33533–33539. [CrossRef] [PubMed]

27. Roos, W.P.; Kaina, B. DNA damage-induced cell death by apoptosis. *Trends Mol. Med.* **2006**, *12*, 440–450. [CrossRef] [PubMed]

28. Cory, S.; Adams, J.M. The BCL2 family: Regulators of the cellular life-or-death switch. *Nat. Rev. Cancer* **2002**, *2*, 647–656. [CrossRef] [PubMed]

29. Van Gurp, M.; Festjens, N.; Van Loo, G.; Saelens, X.; Vandenabeele, P. Mitochondrial intermembrane proteins in cell death. *Biochem. Biophys. Res. Commun.* **2003**, *304*, 487–497. [CrossRef]

30. Zapata, J.M.; Pawlowski, K.; Haas, E.; Ware, C.F.; Godzik, A.; Reed, J.C. A diverse family of proteins containing tumor necrosis factor receptor-associated factor domains. *J. Biol. Chem.* **2001**, *276*, 24242–24252. [CrossRef] [PubMed]

31. Huang, S.T.; Yang, R.C.; Yang, L.H.; Lee, P.N.; Pang, J. Phyllanthus urinaria triggers the apoptosis and Bcl-2 downregulation in Lewis lung carcinoma cells. *Life Sci.* **2003**, *72*, 1705–1716. [CrossRef]

32. Pelicano, H.; Carney, D.; Huang, P. ROS stress in cancer cells and therapeutic implications. *Drug Resist. Updates* **2004**, *7*, 97–110. [CrossRef] [PubMed]

33. Letavayova, L.; Vlckova, V.; Brozmanova, J. Selenium: From cancer prevention to DNA damage. *Toxicology* **2006**, *227*, 1–14. [CrossRef] [PubMed]

34. Turrens, J.F. Mitochondrial formation of reactive oxygen species. *J. Physiol. Lond.* **2003**, *552*, 335–344. [CrossRef] [PubMed]

35. Jaruga, P.; Zastawny, T.H.; Skokowski, J.; Dizdaroglu, M.; Olinski, R. Oxidative DNA base damage and antioxidant enzyme activities in human lung cancer. *FEBS Lett.* **1994**, *341*, 59–64. [CrossRef]

36. Nilsonne, G.; Sun, X.; Nystrom, C.; Rundlof, A.; Fernandes, A.P.; Bjornstedt, M.; Dobra, K. Selenite induces apoptosis in sarcomatoid malignant mesothelioma cells through oxidative stress. *Free Radic. Biol. Med.* **2006**, *41*, 874–885. [CrossRef] [PubMed]

marine drugs

MDPI

Article

Discovery of DNA Topoisomerase I Inhibitors with Low-Cytotoxicity Based on Virtual Screening from Natural Products

Lan-Ting Xin [1,2,†], Lu Liu [1,2,†], Chang-Lun Shao [1,2], Ri-Lei Yu [1,2], Fang-Ling Chen [1,2], Shi-Jun Yue [1,2], Mei Wang [1,2], Zhong-Long Guo [1,2], Ya-Chu Fan [1,2], Hua-Shi Guan [1,2,*] and Chang-Yun Wang [1,2,*]

[1]   Key Laboratory of Marine Drugs, The Ministry of Education of China, School of Medicine and Pharmacy, Ocean University of China, Qingdao 266003, China; xinlanting1993@163.com (L.-T.X.); liu_qd@yahoo.com (L.L.); shaochanglun@163.com (C.-L.S.); rileiyu2010@hotmail.com (R.-L.Y.); Chenfangling0410@hotmail.com (F.-L.C.); shijun_yue@163.com (S.-J.Y.); caoyuxiaowu@163.com (M.W.); 15726227761@163.com (Z.-L.G.); fanyachu@163.com (Y.-C.F.)
[2]   Laboratory for Marine Drugs and Bioproducts, Qingdao National Laboratory for Marine Science and Technology, Qingdao 266071, China
*   Correspondence: hsguan@ouc.edu.cn (H.-S.G.); changyun@ouc.edu.cn (C.-Y.W.); Tel.: +86-532-8203-1667 (H.-S.G.); +86-532-8203-1536 (C.-Y.W.)
†   These authors contributed equally to this work.

Received: 22 April 2017; Accepted: 5 July 2017; Published: 9 July 2017

Abstract: Currently, DNA topoisomerase I (Topo I) inhibitors constitute a family of antitumor agents with demonstrated clinical effects on human malignancies. However, the clinical uses of these agents have been greatly limited due to their severe toxic effects. Therefore, it is urgent to find and develop novel low toxic Topo I inhibitors. In recent years, during our ongoing research on natural antitumor products, a collection of low cytotoxic or non-cytotoxic compounds with various structures were identified from marine invertebrates, plants, and their symbiotic microorganisms. In the present study, new Topo I inhibitors were discovered from low cytotoxic and non-cytotoxic natural products by virtual screening with docking simulations in combination with bioassay test. In total, eight potent Topo I inhibitors were found from 138 low cytotoxic or non-cytotoxic compounds from coral-derived fungi and plants. All of these Topo I inhibitors demonstrated activities against Topo I-mediated relaxation of supercoiled DNA at the concentrations of 5–100 µM. Notably, the flavonoids showed higher Topo I inhibitory activities than other compounds. These newly discovered Topo I inhibitors exhibited structurally diverse and could be considered as a good starting point for the development of new antitumor lead compounds.

Keywords: virtual screening; molecular docking; Topo I inhibitor; low toxic; natural product

1. Introduction

DNA topoisomerase I (Topo I) is a crucial enzyme that works to relax supercoiled DNA during replication, transcription, and mitosis [1,2]. In a number of human solid tumors, the intracellular level of Topo I is higher than that in normal tissues, signifying that controlling the Topo I level is essential in treating cancers [3]. Topo I inhibitors exert their antitumor activities by stabilizing the cleavable Topo I–DNA ternary complex, blocking rejoining of the DNA breaks, and inhibiting enzyme binding to DNA [4,5]. Therefore, Topo I has been considered as a promising target for the development of novel cancer chemotherapeutics [6–8]. Based on the mechanisms of interference with Topo I activity, these Topo I inhibitors can be grouped in two categories: Topo I poisons and Topo I catalytic inhibitors [9].

To date, a large number of Topo-directed agents (e.g., camptothecin (CPT), topotecan, and irinotecan—Figure 1) are known which are currently in clinical use [10,11]. However, their utilities are limited due to the fact that they induce severe toxic side effects such as myelosuppression, nausea, hair loss, congestive heart failure, and in some cases, increase the risk of secondary malignancies [12,13]. Recently, epigallocatechin-3-gallate (EGCG)—a major polyphenolic constituent in green tea—has received much attention as a potential cancer chemopreventive agent with Topo I inhibitory activity (Figure 1) [14–16]. At physiologically attainable concentrations, EGCG exerts growth inhibitory effects on several human tumor cell lines, without affecting normal cell lines, resulting in a dose-dependent inhibition of cell growth [17]. Notably, EGCG possessed low cytotoxicity with much higher half maximal inhibitory concentration (IC_{50}) to human tumor cell lines than the traditional Topo-directed agents [16]. Therefore, low cytotoxic compounds may have the potential with Topo I inhibitory activity and provide the possibility for searching for novel, nontoxic Topo I inhibitors.

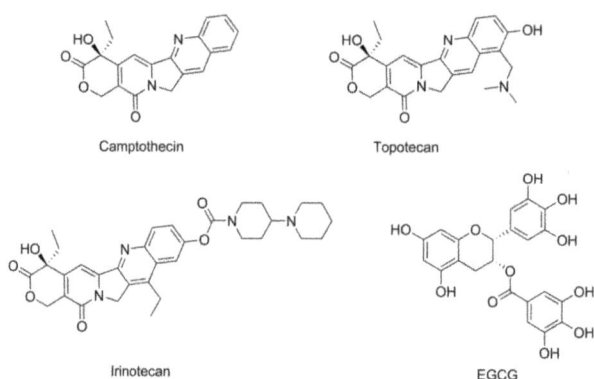

Camptothecin Topotecan

Irinotecan EGCG

Figure 1. Chemical structures of representative DNA topoisomerase I (Topo I) inhibitors.

So far, the discovery of novel Topo I inhibitors has been facilitated by the improvement of a variety of biochemical and cellular assays, as well as molecular docking based on X-ray crystal structures [18–20]. Molecular docking is an application to predict how a protein interacts with small molecules. Based on the docking simulations, virtual screening has become a powerful tool for the discovery of Topo I inhibitors.

In our previous studies, hundreds of antitumor natural products have been isolated from marine invertebrates, plants, and their symbiotic microorganisms [21–24]. During the course of discovering antitumor compounds, a collection of natural products with low cytotoxic or non-cytotoxic activity were also identified. In the present study, from these low cytotoxic and non-cytotoxic natural products, Topo I inhibitors were discovered based on virtual screening with docking simulations in combination with bioassay test. By this approach, eight potent Topo I inhibitors with low cytotoxic or non-cytotoxic activity were found from the natural products isolated from coral-derived fungi and plants.

2. Results and Discussion

In our previous studies, hundreds secondary metabolites were isolated from marine invertebrates, plants, and their symbiotic microorganisms. Among them, there are a number of compounds exhibiting low cytotoxicity or non-cytotoxicity. In this study, 138 compounds (Table S1) from coral-derived fungi and plants with low cytotoxic and non-cytotoxic activity were selected for the screening of Topo I inhibitors by virtual screening combined with bioassay test.

2.1. Virtual Screening

To determine whether the low toxic compounds have potential as Topo I inhibitors, a total of 138 selected compounds were docked into the central catalytic domain of the Topo I–DNA complex (PDB ID: 1K4T) by using molecular operating environment (MOE) program. The docking score at −9.0 kcal/mol was used as a cutoff value for the selection of initial compounds. Thus, the 61 top-ranked complexes were first selected. Then, the selected molecules were further screened based on the following criteria: (1) Complementarity exists between the ligand and the active site of Topo I; (2) Reasonable chemical structures and conformations are in the active site of Topo I. Some unusually highly scored molecules, such as those containing a long aliphatic moiety with many rotatable bonds, were excluded for further evaluation; (3) There is a formation of hydrogen bonds between the ligand and the important residues of Topo I, such as Arg364, Asp533, and Asn722 [25]; (4) The binding mode of the compounds can be reproduced by the LeDock program (cutoff value at −5.0 kcal/mol). As a result, only 27 compounds met the above criteria (Figure 2).

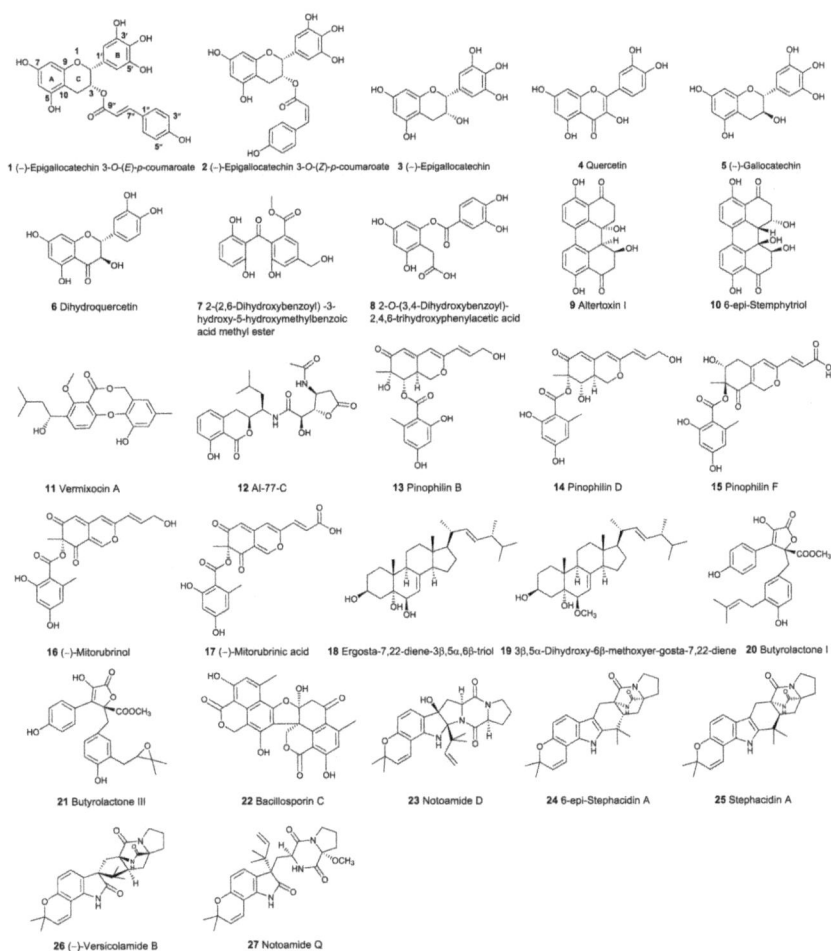

Figure 2. Structures of the potential active compounds from virtual screening.

2.2. DNA Topo I Inhibitory Activity Assay

The above virtual screening results were confirmed by Topo I inhibitory activity assay. Inhibition of the catalytic activity of Topo I has been a useful strategy for the discovery of potential antitumor agents. Topo I creates transient breaks in supercoiled DNA, resulting in DNA relaxation. The relaxed DNA can be distinguished from supercoiled DNA by gel electrophoresis analysis. In the present study, the Topo I inhibitor activities of the selected compounds were detected by monitoring the relaxation of supercoiled DNA by Topo I. Eight of the 27 compounds by virtual screening were discovered to be active against Topo I-mediated relaxation of supercoiled DNA at the concentration of 100 μM (Figure 3). The Topo I inhibition activity of eight hits were further tested at lower concentrations. Among them, four compounds—(−)-epigallocatechin 3-*O*-(*E*)-*p*-coumaroate (**1**), (−)-epigallocatechin 3-*O*-(*Z*)-*p*-coumaroate (**2**), (−)-epigallocatechin (**3**), and quercetin (**4**)—showed activity at 25 μM, and two compounds (**1** and **2**) exhibited activity at 5 μM (Figures 4 and 5). It should be pointed out that compounds **1** and **2** displayed higher inhibitiory activity than EGCG (10 μM) (Figure 6).

In addition, it should be noted that flavonoids showed higher Topo I inhibitory activities than other compounds. The structure–activity relationship (SAR) analysis of these flavonoids revealed that: (1) (−)-epigallocatechin 3-*O*-(*E*)-*p*-coumaroate (**1**) and (−)-epigallocatechin 3-*O*-(*Z*)-*p*-coumaroate (**2**) with a *p*-hydroxy-cinnamic acid group at the position of C-3 could increase the inhibitory activity, while other substitution patterns might not have favorable effects on the activity; (2) the presence of double bonds between C-2 and C-3 as in quercetin (**4**) could enhance the inhibitory activity; (3) the exist of a trihydroxy moiety at the B ring might also improve the activity.

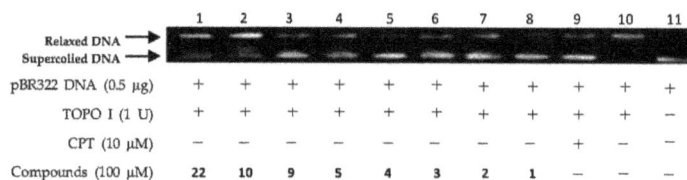

	1	2	3	4	5	6	7	8	9	10	11
pBR322 DNA (0.5 μg)	+	+	+	+	+	+	+	+	+	+	+
TOPO I (1 U)	+	+	+	+	+	+	+	+	+	+	−
CPT (10 μM)	−	−	−	−	−	−	−	−	+	−	−
Compounds (100 μM)	22	10	9	5	4	3	2	1	−	−	−

Figure 3. DNA Topo I inhibitory activities of (−)-epigallocatechin 3-*O*-(*E*)-*p*-coumaroate (**1**), (x)-epigallocatechin 3-*O*-(*Z*)-*p*-coumaroate (**2**), (−)-epigallocatechin (**3**), quercetin (**4**), (−)-gallocatechin (**5**), altertoxin I (**9**), 6-epi-stemphytriol (**10**), and bacillosporin C (**22**) at 100 μM. Lanes 1–8: DNA + Topo I + tested compounds; lane 9: DNA + Topo I + CPT; lane 10: DNA + Topo I; lane 11: DNA.

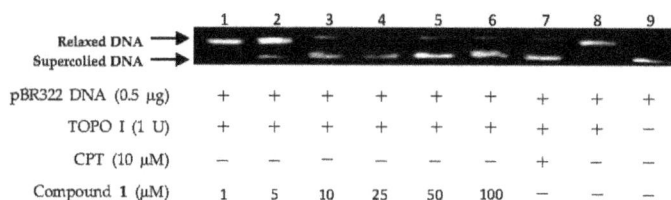

	1	2	3	4	5	6	7	8	9
pBR322 DNA (0.5 μg)	+	+	+	+	+	+	+	+	+
TOPO I (1 U)	+	+	+	+	+	+	+	+	−
CPT (10 μM)	−	−	−	−	−	−	+	−	−
Compound **1** (μM)	1	5	10	25	50	100	−	−	−

Figure 4. DNA Topo I inhibitory activity of (−)-epigallocatechin 3-*O*-(*E*)-*p*-coumaroate (**1**) at various concentrations (1, 5, 10, 25, 50, and 100 μM). Lanes 1–6: DNA + Topo I + compound **1** at various concentrations; lane 7: DNA + Topo I + camptothecin (CPT); lane 8: DNA + Topo I; lane 9: DNA.

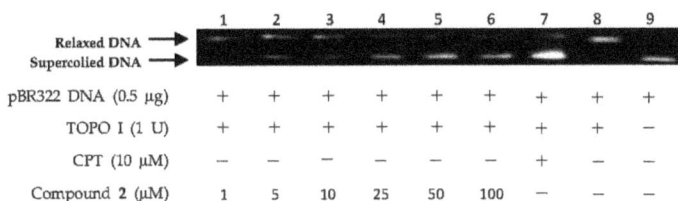

Figure 5. DNA Topo I inhibitory activity of (−)-epigallocatechin 3-*O*-(*Z*)-*p*-coumaroate (**2**) at various concentrations (1, 5, 10, 25, 50 and 100 μM). Lanes 1–6: DNA + Topo I + compound **2** at various concentrations; lane 7: DNA + Topo I + CPT; lane 8: DNA + Topo I; lane 9: DNA.

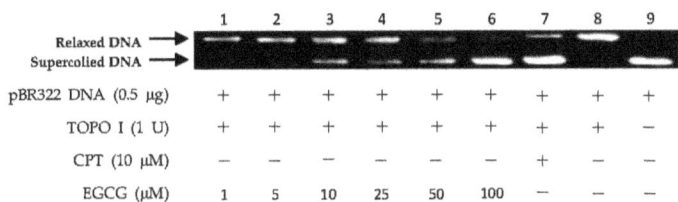

Figure 6. DNA Topo I inhibitory activity of epigallocatechin-3-gallate (EGCG) at various concentrations (1, 5, 10, 25, 50, and 100 μM). Lanes 1–6: DNA + Topo I + EGCG at various concentrations; lane 7: DNA + Topo I + CPT; lane 8: DNA + Topo I; lane 9: DNA.

2.3. Binding Mode of the Representative New Topo I Inhibitors

Among the identified hits, (−)-epigallocatechin 3-*O*-(*E*)-*p*-coumaroate (**1**) and (−)-epigallocatechin 3-*O*-(*Z*)-*p*-coumaroate (**2**)—a pair of isomers—showed the most potent activities in the Topo I inhibition assay. These two compounds have structures similar to that of EGCG, and all of them belong to epigallocatechin. The bonding mode of these representative Topo I inhibitors were observed on the PyMol. In the active site cavity, the orientations of these three compounds were perpendicular to the main axis of the DNA, similar to the known Topo I inhibitors, topotecan, and paralleled to the bases (Figure 7), forming base stacking interactions with the surrounding base pairs (Figure 8). In addition, they could form hydrogen-bonding interactions with the surrounding residues. For example, at the structures of these three flavonoids, the hydroxyl oxygen atoms at the B ring could form hydrogen bonds with the residues of Arg364, Asp533, and Thr718 (Figure 8), resulting in the improvement of binding affinity.

Although the three flavonoids—(−)-epigallocatechin 3-*O*-(*E*)-*p*-coumaroate (**1**), (−)-epigallocatechin 3-*O*-(*Z*)-*p*-coumaroate (**2**), and EGCG—could bind with the Topo I active sites, EGCG shared a different conformation in the active site of Topo I (Figure 8). The benzopyrone moiety of EGCG could deeply intercalate at the DNA cleavage site and stack with the base pairs. Different from EGCG (Figure 8A), in compound **1**, the 5-hydroxyloxygen atom and 7-hydroxyloxygen atom at benzopyrone moiety could form hydrophobic interactions with the surrounding hydrophobic residues, Asn 722 and Arg 488, respectively (Figure 8B). Additionally, the hydrogen bonding interaction was also observed between the 5-hydroxyloxygen atom of compound **2** and Arg364 (Figure 8C). In summary, compounds **1**, **2**, and EGCG have three common features binding to Topo I–DNA complex: (1) a planar aromatic ring could intercalate at the DNA cleavage site; (2) base stacking interactions could form between the ligands and the base pairs; (3) at least three hydrogen bonds could be formed between the ligand and the important residues of Topo I, such as Arg364, Asp533, and Thr718.

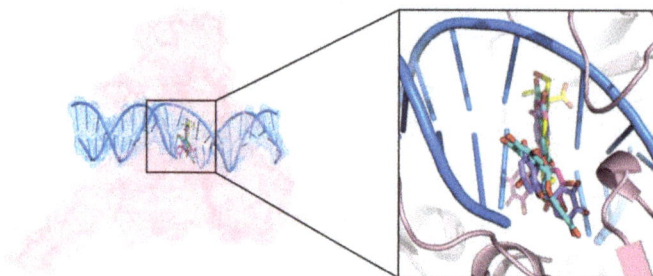

Figure 7. Intercalation of topotecan (yellow), EGCG (pink), (−)-epigallocatechin 3-*O*-(*E*)-*p*-coumaroate (blue), and (−)-epigallocatechin 3-*O*-(*Z*)-*p*-coumaroate (purple) in the Topo I active site.

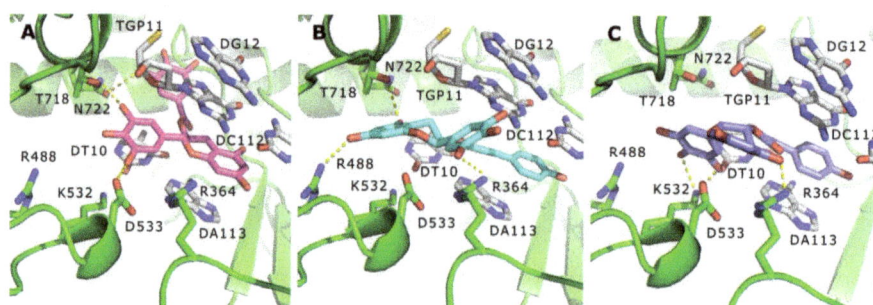

Figure 8. Detailed docked views of different compounds: EGCG (**A**), (−)-epigallocatechin 3-*O*-(*E*)-*p*-coumaroate (**B**), and (−)-epigallocatechin 3-*O*-(*Z*)-*p*-coumaroate (**C**).

3. Materials and Methods

3.1. General Experimental Procedures

Electrophoresis apparatus DYY-8C (Beijing Liuyi Biotechnology Co., Ltd., Beijing, China) was used for electrophoresis analysis. Gel imaging system JS-680B (Shanghai Peiqing Science and Technology Co., Ltd., Shanghai, China) was used for observation of DNA strips. Cooling and heating block CHB-100 (Hangzhou Bioer Technology Co., Ltd., Hangzhou, China) was used for the reaction of DNA and Topo I. Calf thymus Topo I and supercoiled pBR322 plasmid DNA were purchased from Takara Biotechnology Company (Dalian, China). Camptothecin (CPT, 98%) and epigallocatechin-3-gallate (EGCG, 98%) were purchased from Shanghai Aladdin Industrial Corporation (Shanghai, China) and used as positive controls. Dimethyl sulfoxide (DMSO) was purchased from Tianjin Guancheng Chemical Reagent Co., Ltd. (Tianjin, China) and used as solvent.

3.2. Molecular Docking

Molecular docking was used for virtual screening of the selected compounds. The MOE Dock (version 2014. 0901, Chemical Computing Group Inc., Tokyo, Japan) and the LeDock (version 1.0, http://lephar.com/) were operated to dock the compounds into the active sites of the known antitumor target protein. The molecular mechanic force field of MOE Dock used in molecular docking was set as AMBER10: EHT. Two rescoring functions, including London dG and GBVI/WSA dG, were used for pose scoring [26].

The X-ray crystallographic of Topo I–DNA in complex with topotecan (PDB ID: 1K4T) was downloaded from the Protein Data Bank (PDB, http://www.rcsb.org/) and used as a reference model for molecular docking [19]. Suitable protonation of the protein was executed at physiological pH. By using MOE software (version 2014. 0901, Chemical Computing Group Inc, Tokyo, Japan) at the AMBER10: EHT force field, water molecules were removed, hydrogen atoms were added, and energy was minimized [27].

The structures of the ligands were generated in the cdx format using the ChemBio Draw Ultra (version 14.0, PerkinElmer Inc., Fremont, CA, USA). These ligands were converted to the mol2 format and the structures were optimized by the function of minimize energy in ChemBio 3D Ultra (version 14.0, PerkinElmer Inc., Fremont, CA, USA). Further optimizations of the structures of these molecules were made by the energy minimization module in MOE software. During energy minimization by MOE software, the AMBER10 force field was used. Energy minimization was converged when the energy gradient reached 0.01 kcal/mol/Å^3. All of the ligands used for the docking studies were assigned to suitable protonation status corresponding to physiological pH [27].

3.3. Preparation of the Tested Compounds

From the secondary metabolites isolated from marine invertebrates, plants, and their symbiotic microorganisms in our lab, 138 compounds (Table S1) with low cytotoxic or non-cytotoxic activity isolated from coral-derived fungi and plants were selected for virtual screening and DNA Topo I inhibition assay. For the bioassay, the tested compounds were firstly dissolved in DMSO, and then diluted with DMSO to obtain a serial solution with the concentrations of 100, 50, 25, 10, 5, and 1 μM. EGCG was also dissolved with the concentrations of 100, 50, 25, 10, 5, and 1 μM. The positive control CPT was prepared at the concentration of 10 μM.

3.4. DNA Topoisomerase I Inhibitory Activity Assay

The Topo I inhibitory activity was measured by assessing the relaxation of supercoiled pBR322 plasmid DNA. The reaction mixture (20 μL each), containing 35 mM Tris-HCI (pH 8.0), 72 mM KCI, 5 mM MgCl$_2$, 5 mM dithiothreitol (DTT), 5 mM spermidine, 0.01% bovine serum albumin (BSA), 0.5 μg pBR322 plasmid DNA, 1.0 U calf thymus DNA Topo I, and 0.2 μL various concentrations of tested compounds, were incubated at 37 °C for 30 min. The reactions were terminated by adding dye solution containing 1% SDS, 0.02% bromophenol blue, and 50% glycerol. The mixtures were applied to 1% agarose gel and subjected to electrophoresis for 1 h in Tris-borate-EDTA buffer (0.089 mM). The gel was stained with Gelred and visualized under UV illumination and then photographed with a Gel imaging system.

4. Conclusions

In the present study, Topo I inhibitors were discovered from natural products with low cytotoxic and non-cytotoxic activity by virtual screening with docking simulations in combination with bioassay test. The 27 compounds with potential Topo I inhibition activity were screened from 138 marine and plant-derived natural products by means of virtual screening. On the basis of virtual screening, eight active compounds were discovered through the verification approach by bioassay. All of these Topo I inhibitors were found to be active against Topo I-mediated relaxation of supercoiled DNA at the concentrations of 5–100 μM. Notably, the flavonoids showed higher Topo I inhibitory activities than other compounds. The above results suggested that the low cytotoxic or non-cytotoxic compounds might possess Topo I inhibitory activities, and have value to be further studied for the rational drug design of antitumor agents.

Supplementary Materials: The following are available online at http://www.mdpi.com/1660-3397/15/7/217/s1. Table S1: 138 compounds derived from coral-derived fungi and plants; Table S2: The binding energy of 138 compounds bound with the crystal structure of the ternary complex of topotecan-DNA-Topo I (PDB ID: 1K4T); Figure S1: DNA Topo I inhibitory activities of (−)-epigallocatechin 3-O-(E)-p-coumaroate (**1**), (−)-epigallocatechin 3-O-(Z)-p-coumaroate (**2**), (−)-epigallocatechin (**3**), quercetin (**4**) and altertoxin I (**9**) at 50 μM; Figure S2: DNA Topo I inhibitory activities of (−)-epigallocatechin 3-O-(E)-p-coumaroate (**1**), (−)-epigallocatechin 3-O-(Z)-p-coumaroate (**2**), (−)-epigallocatechin (**3**), quercetin (**4**) and altertoxin I (**9**) at

25 µM; Figure S3: DNA Topo I inhibitory activities of (−)-epigallocatechin 3-*O*-(*E*)-*p*-coumaroate (**1**) and (−)-epigallocatechin 3-*O*-(*Z*)-*p*-coumaroate (**2**) at 10, 5 and 1 µM; Figure S4: DNA Topo I inhibitory activities of (−)-epigallocatechin (**3**) and quercetin (**4**) at 10, 5 and 1 µM.

Acknowledgments: We would like to thank Jing-Shuai Wu (Ocean University of China) and Ting Shi (Ocean University of China) for critically reading a previous version of this manuscript. This work was supported by the National High Technology Research and Development Program of China (863 Program) (No. 2013AA093001), The Scientific and Technological Innovation Project Financially Supported by Qingdao National Laboratory for Marine Science and Technology (No. 2015ASKJ02), and the Taishan Scholars Program, China.

Author Contributions: C.-Y.W. and H.S.G. conceived of and proposed the idea; L.-T.X. and L.L. designed the study; L.-T.X., L.L., F.-L.C. S.-J.Y., M.W., Z.-L.G. and Y.-C.F. performed the experiments; L.-T.X., L.L., C.-L.S. R.-L.Y., F.-L.C. and S.-J.Y. participated in data analysis; C.-Y.W., L.L. and S.-J.Y. contributed to writing, revising and proof-reading the manuscript. All authors read and approved the final manuscript.

Conflicts of Interest: The authors declare no conflict of interest.

References

1. Redinbo, M.R.; Stewart, L.; Kuhn, P.; Champoux, J.J.; Hol, W.G.J. Crystal structures of human topoisomerase I in covalent and noncovalent complexes with DNA. *Science* **1998**, *279*, 1504–1513. [CrossRef] [PubMed]
2. Wang, J.C. DNA topoisomerases. *Annu. Rev. Biochem.* **1996**, *65*, 635–692. [CrossRef] [PubMed]
3. Liew, S.T.; Yang, L.X. Design, synthesis and development of novel camptothecin drugs. *Curr. Pharm. Des.* **2008**, *14*, 1078–1097. [CrossRef] [PubMed]
4. Pommier, Y. DNA topoisomerase I inhibitors: Chemistry, biology, and interfacial inhibition. *Chem. Rev.* **2009**, *109*, 2894–2902. [CrossRef] [PubMed]
5. Pommier, Y.; Tanizawa, A.; Kohn, K.W. Mechanisms of topoisomerase I inhibition by anticancer drugs. *Adv. Pharmacol.* **1994**, *29*, 73–92.
6. Takagi, K.; Dexheimer, T.S.; Redon, C.; Sordet, O.; Agama, K.; Lavielle, G.; Pierre, A.; Bates, S.E.; Pommier, Y. Novel E-ring camptothecin keto analogues (S38809 and S39625) are stable, potent, and selective topoisomerase I inhibitors without being substrates of drug efflux transporters. *Mol. Cancer Ther.* **2007**, *6*, 3229–3238. [CrossRef] [PubMed]
7. Bjornsti, M.A.; Benedetti, P.; Viglianti, G.A.; Wang, J.C. Expression of human DNA topoisomerase I in yeast cells lacking yeast DNA topoisomerase I: Restoration of sensitivity of the cells to the antitumor drug camptothecin. *Cancer Res.* **1989**, *49*, 6318–6323. [PubMed]
8. Pommier, Y.; Pourquier, P.; Urasaki, Y.; Wu, J.; Laco, G.S. Topoisomerase I inhibitors: Selectivity and cellular resistance. *Drug Resist. Updates* **1999**, *2*, 307–318. [CrossRef] [PubMed]
9. Pommier, Y. Diversity of DNA topoisomerases I and inhibitors. *Biochimie* **1998**, *80*, 255–270. [CrossRef]
10. Rasheed, Z.A.; Rubin, E.H. Mechanisms of resistance to topoisomerase I-targeting drugs. *Oncogene* **2003**, *22*, 7296–7304. [CrossRef] [PubMed]
11. Li, Q.Y.; Zu, Y.G.; Shi, R.Z.; Yao, L.P. Review camptothecin: Current perspectives. *Curr. Med. Chem.* **2006**, *13*, 2021–2039. [CrossRef] [PubMed]
12. Chen, A.Y.; Liu, L.F. DNA topoisomerases: Essential enzymes and lethal targets. *Annu. Rev. Pharmacol. Toxicol.* **1994**, *34*, 191–218. [CrossRef] [PubMed]
13. Anderson, R.D.; Berger, N.A. Mutagenicity and carcinogenicity of topoisomerase-interactive agents. *Mutat. Res. Fund. Mol. Mech.* **1994**, *309*, 109–142. [CrossRef]
14. Landis-Piwowar, K.R.; Kuhn, D.J.; Wan, S.B.; Chen, D.; Chan, T.H.; Dou, Q.P. Evaluation of proteasome-inhibitory and apoptosis-inducing potencies of novel (−)-EGCG analogs and their prodrugs. *Int. J. Mol. Med.* **2005**, *15*, 735–742. [CrossRef] [PubMed]
15. Tachibana, H.; Koga, K.; Fujimura, Y.; Yamada, K. A receptor for green tea polyphenol EGCG. *Nat. Struct. Mol. Biol.* **2004**, *11*, 380–381. [CrossRef] [PubMed]
16. Berger, S.J.; Gupta, S.; Belfi, C.A.; Gosky, D.M.; Mukhtar, H. Green tea constituent (−)-epigallocatechin-3-gallate inhibits topoisomerase I activity in human colon carcinoma cells. *Biochem. Bioph. Res. Commun.* **2001**, *288*, 101–105. [CrossRef] [PubMed]
17. Ahmad, N.; Feyes, D.K.; Agarwal, R.; Mukhtar, H.; Nieminen, A.L. Green tea constituent epigallocatechin-3-gallate and induction of apoptosis and cell cycle arrest in human carcinoma cells. *J. Natl. Cancer Inst.* **1997**, *89*, 1881–1886. [CrossRef] [PubMed]

18. Dong, G.; Sheng, C.; Wang, S.; Miao, Z.; Yao, J.; Zhang, W. Selection of evodiamine as a novel topoisomerase I inhibitor by structure-based virtual screening and hit optimization of evodiamine derivatives as antitumor agents. *J. Med. Chem.* **2010**, *53*, 7521–7531. [CrossRef] [PubMed]

19. Zhuo, S.T.; Li, C.Y.; Hu, M.H.; Chen, S.B.; Yao, P.F.; Huang, S.L.; Ou, T.M.; Tan, J.H.; An, L.K.; Li, D.; et al. Synthesis and biological evaluation of benzo [*a*] phenazine derivatives as a dual inhibitor of topoisomerase I and II. *Org. Biomol. Chem.* **2013**, *11*, 3989–4005. [CrossRef] [PubMed]

20. Song, Y.; Shao, Z.; Dexheimer, T.S.; Scher, E.S.; Pommier, Y.; Cushman, M. Structure-Based Design, Synthesis and Biological Studies of New Anticancer Norindenoisoquinoline Topoisomerase I Inhibitors. *J. Med. Chem.* **2010**, *53*, 1979–1989. [CrossRef] [PubMed]

21. Zheng, C.J.; Shao, C.L.; Guo, Z.Y.; Chen, J.F.; Deng, D.S.; Yang, K.L.; Chen, Y.Y.; Fu, X.M.; She, Z.G.; Lin, Y.C.; et al. Bioactive hydroanthraquinones and anthraquinone dimers from a soft coral-derived *Alternaria* sp. fungus. *J. Nat. Prod.* **2012**, *75*, 189–197. [CrossRef] [PubMed]

22. Cao, F.; Wu, Z.H.; Shao, C.L.; Pang, S.; Liang, X.Y.; Voogd, N.J.D.; Wang, C.Y. Cytotoxic scalarane sesterterpenoids from the South China Sea sponge *Carteriospongia Foliascens*. *Org. Biomol. Chem.* **2015**, *13*, 4016–4024. [CrossRef] [PubMed]

23. Chen, M.; Wu, X.D.; Zhao, Q.; Wang, C.Y. Topsensterols A–C, Cytotoxic Polyhydroxylated Sterol Derivatives from a Marine Sponge *Topsentia* sp. *Mar. Drugs* **2016**, *14*, 146. [CrossRef] [PubMed]

24. Zhao, D.L.; Shao, C.L.; Gan, L.S.; Wang, M.; Wang, C.Y. Chromone derivatives from a sponge-derived strain of the fungus *Corynespora cassiicola*. *J. Nat. Prod.* **2015**, *78*, 286–293. [CrossRef] [PubMed]

25. Staker, B.L.; Feese, M.D.; Cushman, M.; Pommier, Y.; Zembower, D.; Stewart, L.; Burgin, A.B. Structures of three classes of anticancer agents bound to the human topoisomerase I-DNA covalent complex. *J. Med. Chem.* **2005**, *48*, 2336–2345. [CrossRef] [PubMed]

26. Liu, M.S.; Zheng, N.; Li, D.M.; Zheng, H.L.; Zhang, L.L.; Ge, H.; Liu, W.D. *cyp51A*-based mechanism of azole resistance in *Aspergillus fumigatus*: Illustration by a new 3D Structural Model of *Aspergillus fumigatus* CYP51A protein. *Med. Mycol.* **2016**, *54*, 400–408. [CrossRef] [PubMed]

27. Mamidala, R.; Majumdar, P.; Jha, K.K.; Bathula, C.; Agarwal, R.; Chary, M.T.; Mazumdar, H.K.; Munshi, P.; Sen, S. Identification of *Leishmania donovani* Topoisomerase 1 inhibitors via intuitive scaffold hopping and bioisosteric modification of known Top 1 inhibitors. *Sci. Rep.* **2016**, *6*, 26003.

![marine drugs logo] *marine drugs*

Article

Fucoidan Does Not Exert Anti-Tumorigenic Effects on Uveal Melanoma Cell Lines

Michaela Dithmer [1], Anna-Maria Kirsch [1], Elisabeth Richert [1], Sabine Fuchs [2], Fanlu Wang [2], Harald Schmidt [3], Sarah E. Coupland [4], Johann Roider [1] and Alexa Klettner [1,*]

[1] Department of Ophthalmology, University of Kiel, University Medical Center, Arnold-Heller-Str. 3, 24105 Kiel, Germany; michaela_dithmer@web.de (M.D.); anna.maria.kirsch@gmx.de (A.-M.K.); elisabeth.richert@uksh.de (E.R.); johann.roider@uksh.de (J.R.)

[2] Experimental Trauma Surgery, University of Kiel, University Medical Center, Arnold-Heller-Str. 3, 24105 Kiel, Germany; sabine.fuchs@uksh.de (S.F.); fanlu.wang@uksh.de (F.W.)

[3] MetaPhysiol, Am Römerberg, 55270 Essenheim, Germany; schmidt@metaphysiol.de

[4] Department of Molecular and Clinical Cancer Medicine, Liverpool Ocular Oncology Research Group, Pathology, University of Liverpool, Liverpool L69 3BX, UK; s.e.coupland@liverpool.ac.uk

* Correspondence: aklettner@auge.uni-kiel.de; Tel.: +49-431-500-24283; Fax: +49-431-500-24244

Received: 26 January 2017; Accepted: 18 June 2017; Published: 22 June 2017

Abstract: Background. The polysaccharide fucoidan is widely investigated as an anti-cancer agent. Here, we tested the effect of fucoidan on uveal melanoma cell lines. Methods. The effect of 100 μM fucoidan was investigated on five cell lines (92.1, Mel270 OMM1, OMM2.3, OMM2.5) and of 1 μg/mL–1 mg/mL fucoidan in two cell lines (OMM1, OMM2.3). Cell proliferation and viability were investigated with a WST-1 assay, migration in a wound healing (scratch) assay. Vascular Endothelial Growth Factor (VEGF) was measured in ELISA. Angiogenesis was evaluated in co-cultures with endothelial cells. Cell toxicity was induced by hydrogen-peroxide. Protein expression (Akt, ERK1/2, Bcl-2, Bax) was investigated in Western blot. Results. Fucoidan increased proliferation in two and reduced it in one cell line. Migration was reduced in three cell lines. The effect of fucoidan on VEGF was cell type and concentration dependent. In endothelial co-culture with 92.1, fucoidan significantly increased tubular structures. Moreover, fucoidan significantly protected all tested uveal melanoma cell lines from hydrogen-peroxide induced cell death. Under oxidative stress, fucoidan did not alter the expression of Bcl-2, Bax or ERK1/2, while inducing Akt expression in 92.1 cells but not in any other cell line. Conclusion. Fucoidan did not show anti-tumorigenic effects but displayed protective and pro-angiogenic properties, rendering fucoidan unsuitable as a potential new drug for the treatment of uveal melanoma.

Keywords: fucoidan; uveal melanoma; VEGF; angiogenesis; oxidative stress

1. Introduction

Uveal melanoma (UM) is the most common primary tumor of the adult eye with an incidence of 4–8 per million in Western countries [1]. It arises from melanocytes of the uvea, the tissue between the inner retina and the outer scleral layer of the posterior eye, including the iris, ciliary body and choroid. Most UM arise from the choroid, which provides blood supply and maintenance for the photoreceptors of the retina. The disease generally occurs in the 6th decade of life and primarily affects fair-skinned people of Caucasian descent [2]. Treatment options for UM depend on the tumor size and patient choice, but include transpupillary thermotherapy, radiation therapy (including plaque brachytherapy, proton beam- and gamma-knife radiotherapy), local tumor resection and enucleation [2]. Radiation therapy is conducted with good success for medium sized tumors, however, it may result in profound vision loss due to side effects [3]. Metastases develop in up to 50% of UM patients, primarily affecting the liver. The prognosis of these patients is poor, as the current treatment options for metastatic UM are very limited [1,4]. New treatment options for this disease are currently of great interest and are an important activity of numerous basic and clinical research teams.

A promising new approach in the treatment of cancer is the use of fucoidan, a sulfated polysaccharide, obtained from the cell-wall matrix of brown algae. Fucoidan contains high amounts of L-fucose, but has a highly complex structure and may differ substantially depending on different species, regional origin and even mode of extraction [5]. Fucoidan has been reported in several studies to have anti-tumorigenic properties, e.g., it has been shown to be anti-proliferative and/or pro-apoptotic on several kind of tumors cells, such as colon cancer [6], hepatoma [7], urinary bladder cancer cells [8], breast cancer [9], melanoma cells [10] or prostate cancer cells [11]. Fucoidan has also shown anti-angiogenic properties [6,12,13] and is discussed as a promising anti-cancer agent [14]. Therefore, fucoidan might be an interesting new therapeutic compound for the treatment of UM.

Important parameters in tumor progression are proliferation, migration and angiogenic potential [5]. We tested the effect of fucoidan on these parameters in five different UM cell lines. One of the factors that have been discussed to be involved in the pathogenesis of UM is Vascular Endothelial Growth Factor (VEGF). VEGF has been reported in UM, ocular fluid of UM patients and UM cell lines [15–17]. A meta-analysis showed that VEGF expression in patients with UM was significantly higher compared to controls [18]. Moreover, VEGF has been elevated in patients with metastatic UM [17], and has been proposed to be a marker for high risk patients [18]. Fucoidan has been reported to reduce VEGF expression in breast cancer cells [9] and in Lewis tumor bearing mice [19]. Therefore, we investigated the effect of fucoidan on VEGF secretion by UM cells.

Oxidative stress is an important factor in tumor pathology and metastasis [20,21] and is utilized by therapeutic compounds to destroy the tumor tissue [22]. In primary UM, the tumor is treated with ionizing radiation, which induces cell death via oxidative stress-mediated killing of tumor cells [3,23]. Therefore, we also tested the effect of fucoidan on UM cells stressed with Hydrogen peroxide (H_2O_2). Fucoidan has been shown to exert its anti-tumor functions via ERK1/2, Akt [6,11,12], Bcl-2 and Bax [8,9,24,25]; all these proteins have also been implicated in the pathogenesis of UM [26–31]. Therefore, we also assessed how fucoidan affects the expression of these proteins under oxidative stress.

2. Results

2.1. Proliferation

Fucoidan had a cell specific effect on cell proliferation. In 92.1 cells, fucoidan induced a significant increase in cell number one day ($p < 0.05$), two days ($p < 0.01$) and three days ($p < 0.05$) after incubation, while in Mel270 cells, fucoidan reduced proliferation after two and three days (both $p < 0.05$). OMM1 and OMM2.3 were not affected by fucoidan, while in OMM2.5 cells, fucoidan increased cell number significantly after one day of incubation ($p < 0.001$) (Figure 1). In addition, for OMM1 and OMM2.3, different concentrations (1 µg/mL, 10 µg/mL, 100 µg/mL, 1 mg/mL) after one day of incubation were tested. Fucoidan did not show any significant effect in either cell line or in either concentration (Figure 2).

2.2. Wound Healing/Migration

Fucoidan induced a significant decrease in wound healing ability in 92.1 cells, OMM2.3, and OMM2.5 cells (all $p < 0.05$). No significant effect was seen on Mel270 and OMM1 cells (Figure 3).

2.3. VEGF Secretion

We have previously shown that all tested UM cell lines secrete VEGF [32] and that this batch of fucoidan reduces VEGF in retinal pigment epithelial cells in the tested concentration [33]. Fucoidan (100 µg/mL) did not inhibit VEGF secretion in any of the UM cell lines when incubated for up to three days (Figure 4). However, these results are dose and cell-line dependent. In a separate set of experiments, we investigated different concentrations of fucoidan (1 µg/mL, 10 µg/mL, 100 µg/mL, 1 mg/mL) in OMM1 and OMM2.3 cells after treatment for one day. While for OMM2.3 cells, a slight but significant induction of VEGF could be found at 10 and 100 µg/mL, fucoidan at 1 mg/mL significantly reduced VEGF in OMM1 cells (Figure 5).

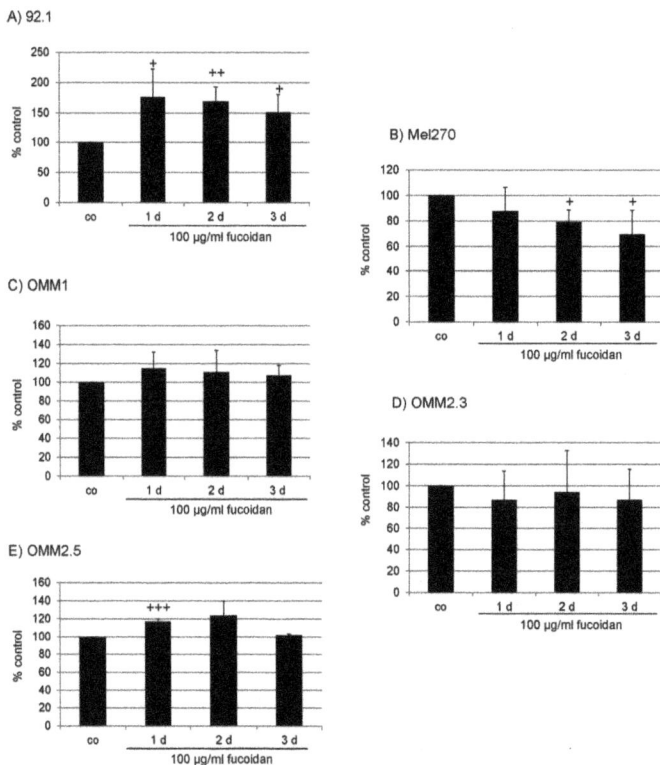

Figure 1. Proliferation (time line). Proliferation of uveal melanoma cells was tested after incubation with fucoidan (100 µg/mL) for one, two, and three days in (**A**) 92.1; (**B**) Mel 270; (**C**) OMM1; (**D**) OMM2.3 and (**E**) OMM2.5 cells. Fucoidan exhibited a cell specific effect with an acceleration of proliferation in 92.1 and OMM2.5 cells, but a decrease in Mel270 cells. Statistical significance was evaluated with student's t-test. $+ p < 0.05$ compared to control, $++ p < 0.01$ compared to control, $+++ p < 0.001$ compared to control. Co: control.

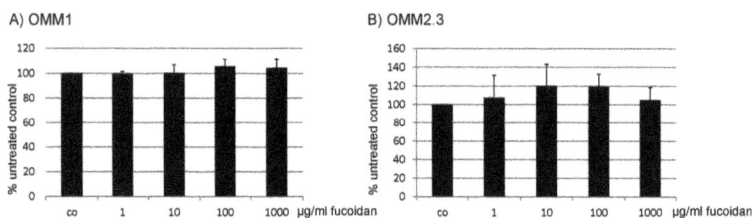

Figure 2. Proliferation (concentration). Proliferation of uveal melanoma cell lines (**A**) OMM1 and (**B**) OMM2.3 was tested after one day of treatment with 1 µg/mL, 10 µg/mL, 100 µg/mL or 1000 µg/mL fucoidan. No significant effect on proliferation was found. Statistical significance was evaluated with student's t-test.

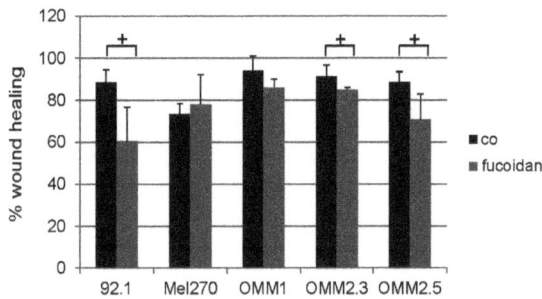

Figure 3. Wound healing. Wound healing ability of uveal melanoma cells was tested after incubation with fucoidan (100 µg/mL) for one day in 92.1, Mel 270, OMM1, OMM2.3 and OMM2.5 cells. Fucoidan significantly decreased wound healing in 92.1, OMM2.3 and OMM2.5 cells. Statistical significance was evaluated with student's *t*-test. $+ p < 0.05$ compared to control. Co = control.

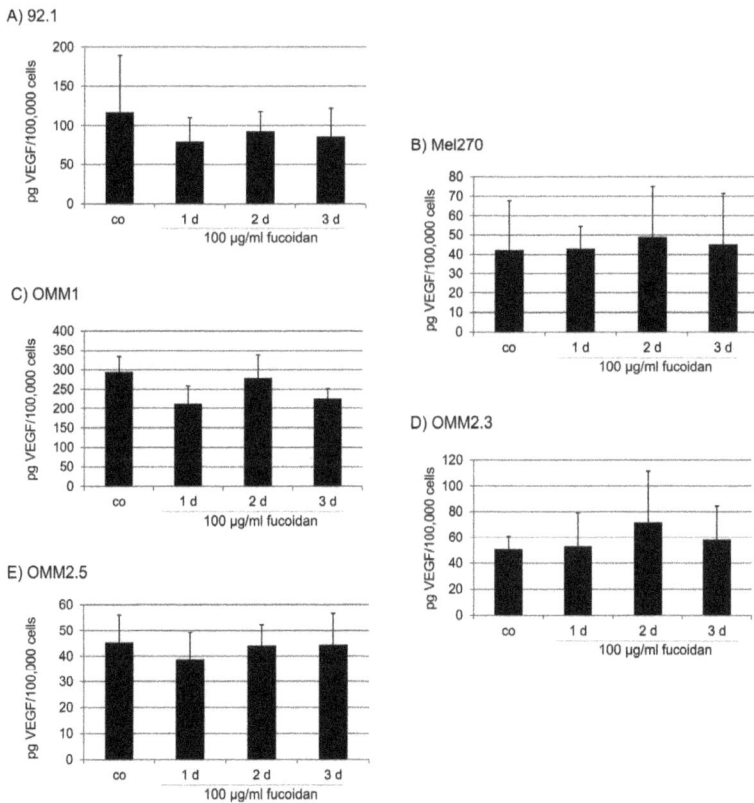

Figure 4. Vascular Endothelial Growth Factor (VEGF) secretion (time line). Influence of fucoidan (100 µg/mL) on VEGF secretion by uveal melanoma cell lines. Treatment with fucoidan for up to three days did not show any significant influence on the secretion of VEGF in any of the cell lines tested (**A**) 92.1; (**B**) Mel270; (**C**) OMM1; (**D**) OMM2.3; (**E**) OMM2.5. The secretion of VEGF was determined in VEGF-ELISA. Statistical significance was evaluated with student's *t*-test. Co: control.

A) OMM1   B) OMM2.3

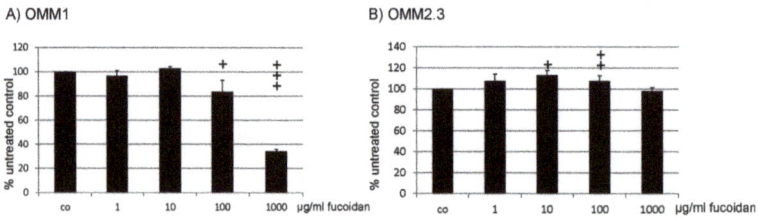

Figure 5. VEGF secretion (concentration). Influence of different concentrations of fucoidan (1 µg/mL–1 mg/mL) on VEGF secretion in (**A**) OMM1 and (**B**) OMM2.3 uveal melanoma cell lines. Treatment with fucoidan displayed a dose- and cell-dependent effect with significant reduction of VEGF in OMM1 (100 µg/mL, 1 mg/mL) and a slight but significant induction in OMM2.3 cells (10 µg/mL, 100 µg/mL). The secretion of VEGF was determined in VEGF-ELISA. Statistical significance was evaluated with student's *t*-test. $+ p < 0.05$, $++ p < 0.01$, $+++ p < 0.001$. Co: control.

2.4. Angiogenesis

Fucoidan induced an elevation of the tubular area in a co-culture of endothelial cells with 92.1 cells ($p < 0.01$). Similarly, fucoidan increased tubular length ($p < 0.01$). Fucoidan did not, however, influence the total area of endothelial coverage in these co-cultures. No effect was seen in co-cultures of endothelial cells with the metastatic UM cell line OMM2.3 (Figure 6).

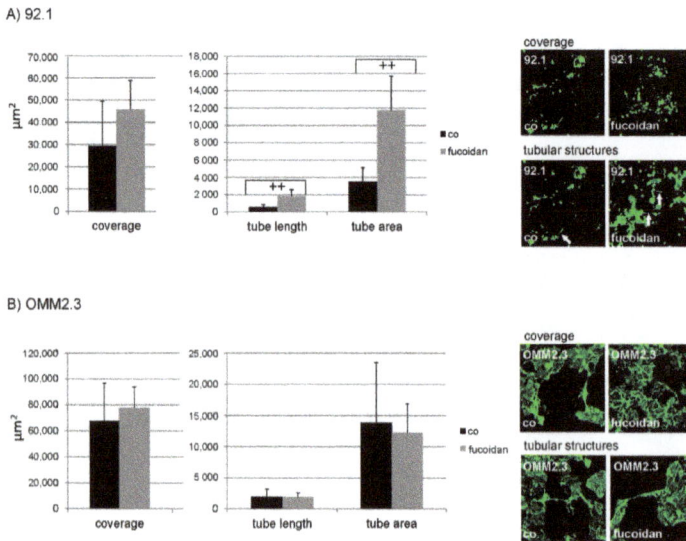

Figure 6. Tubular structures in endothelial–uveal melanoma cell line co-culture. Uveal melanoma cell line 92.1 and OMM2.3 were co-cultured with outgrowth endothelial cells and subjected to 100 µg/mL fucoidan. In co-cultures with endothelial cells and 92.1 cell line (**A**), tubular area and tubular length were increased by fucoidan. Total coverage with endothelial cells, however, was not influenced. In co-cultures with endothelial cells and OMM2.3 cell line, fucoidan displayed no effect (**B**). Statistical significance was evaluated with student's *t*-test. $++ p < 0.01$ compared to control. Co = control.

2.5. Protection

We have previously shown that the UM cell lines have a different susceptibility towards H$_2$O$_2$-induced cell toxicity [32]. In all cell lines tested, fucoidan exerted a significant protection on UM cell lines under oxidative stress. (92.1, 250 μM H_2O_2, $p < 0.01$; Mel270, 500 μM H_2O_2, $p < 0.01$; OMM1, 500 μM H_2O_2, $p < 0.05$; OMM2.3, 1000 μM H_2O_2, $p < 0.001$; OMM2.5, 1000 μM H_2O_2, $p < 0.001$) (Figure 7).

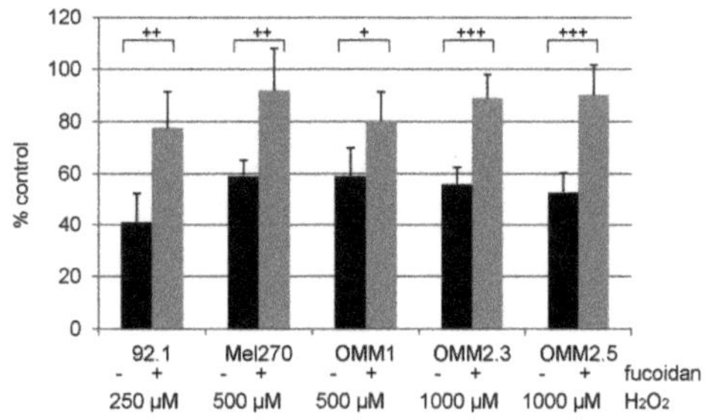

Figure 7. Cell viability of uveal melanoma cell lines under oxidative stress. Uveal melanoma cell lines 92.1, Mel270, OMM1, OMM2.3, and OMM2.5 were subjected to 250 μM (92.1), 500 μM, (Mel270 and OMM1) or 1000 μM (OMM2.3 and OMM2.5) H_2O_2. The toxicity of these concentrations of H_2O_2 in the respective cell line has been shown previously [32]. The ability of 100 μg/mL fucoidan to protect cell viability after H_2O_2 treatment was detected in WST assay. All tested substances exhibited statistically significant protection in all cell lines tested. Statistical significance was evaluated with student's *t*-test. $+ p < 0.05$, $++ p < 0.01$, $+++ p < 0.001$.

2.6. Protein Expression

Under oxidative stress conditions fucoidan did not show any influence on Bcl-2 or Bax expression in any of the cell lines (Figure 8). In 92.1 cell lines, fucoidan induced a significant induction of Akt expression compared to cells treated with H_2O_2 alone, while it showed no significant effect on the other cell lines (Figure 9). Considering ERK1/2, no statistically significant change in ERK1/2 expression or phosphorylation compared to H_2O_2-treated cells can be found (Figure 10).

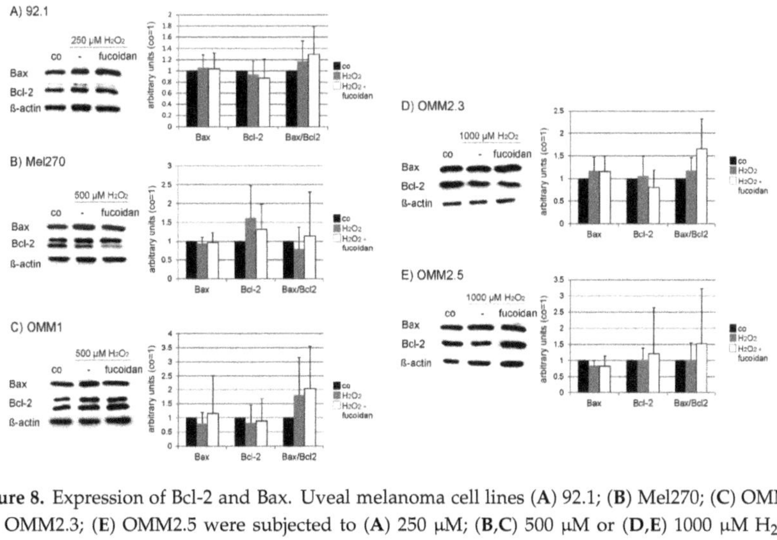

Figure 8. Expression of Bcl-2 and Bax. Uveal melanoma cell lines (**A**) 92.1; (**B**) Mel270; (**C**) OMM1; (**D**) OMM2.3; (**E**) OMM2.5 were subjected to (**A**) 250 μM; (**B,C**) 500 μM or (**D,E**) 1000 μM H_2O_2. The effect of 100 μg/mL fucoidan on the expression of Bcl-2 and Bax was investigated in Western blot. Example blots (compound) and densitometric evaluations are shown. Statistical significance was evaluated with student's *t*-test.

A) Densitometric Evaluation

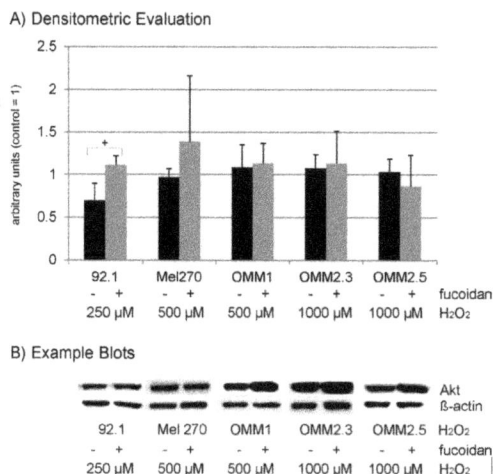

Figure 9. Expression of Akt. Uveal melanoma cell lines 92.1, Mel270, OMM1, OMM2.3 and OMM2.5 were subjected to 250 μM (92.1), 500 μM (Mel270 and OMM1) or 1000 μM (OMM2.3 and OMM2.5) H_2O_2. The effect 100 μg/mL fucoidan on the expression of Akt was investigated in Western blot. Densitometric evaluations (**A**) and example blots (compound) (**B**) are shown. Statistical significance was evaluated with student's *t*-test. $+ p < 0.05$.

A) Densitometric Evaluation ERK(42 kDa)

B) Densitometric Evaluation pERK(42 kDa)

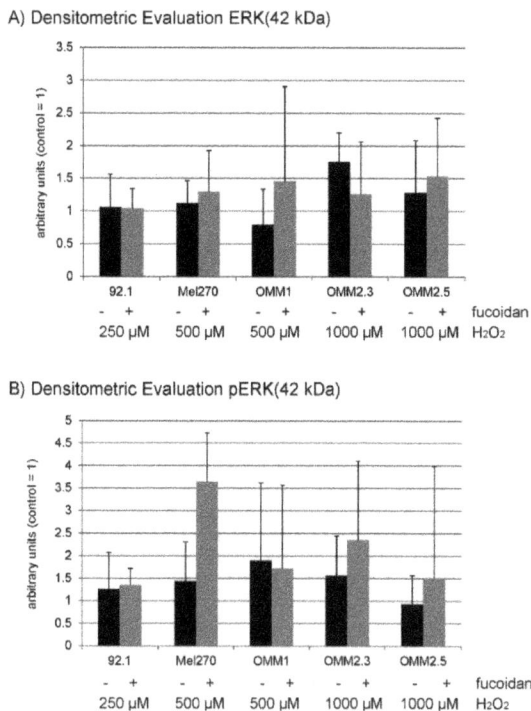

Figure 10. *Cont.*

C) Example Blots

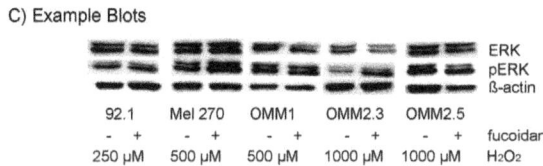

Figure 10. Expression and phosphorylation of ERK1/2. Uveal melanoma cell lines 92.1, Mel270, OMM1, OMM2.3 and OMM2.5 were subjected to 250 μM (92.1), 500 μM (Mel270 and OMM1) or 1000 μM (OMM2.3 and OMM2.5) H_2O_2. The effect of 100 μg/mL fucoidan on the expression and phosphorylation of ERK1/2 was investigated in Western blot. Densitometric evaluation of (**A**) ERK and (**B**) pERK blots are shown for the 42 kDa isoform. (**C**) Example blots (compound). Statistical significance was evaluated with student's *t*-test.

3. Discussion

Fucoidan has been shown to display a variety of anti-tumor effects on several types of tumors or cancer cell lines. Here, we investigated its effect on UM, a primary malignant neoplasm of the eye. We investigated classical parameters, such as proliferation, migration, VEGF secretion and angiogenesis and additionally investigated the effect of fucoidan on H_2O_2-induced cell death and protein expression.

Anti-proliferative activity of fucoidan has been shown for several cancer cell types, such as bronchopulmonary carcinoma [34], cutaneous melanoma cells [10,35], bladder cancer cells [8], breast cancer cells [9], or B-cell lymphoma [36]. In our study, fucoidan reduced proliferation in one cell line (Mel 270), but, surprisingly, it induced proliferation in two cell lines (92.1 and OMM2.5). In OMM1 and OMM2.3 cells, both time line (100 μg/mL) and concentration (1 μg/mL–1 mg/mL) was tested and no effect on proliferation was seen at either time point or contranction. The effect, therefore, is clearly cell type specific. Moreover, the pro-proliferative effect on two cell lines would be a worrisome result if fucoidan were to be used in UM patients.

Fucoidan decreased its wound healing ability in three (92.1; OMM2.3; OMM2.5) out of five UM cell lines. This indicates that fucoidan interferes with migration in these cell lines, especially as wound healing assay measure both proliferation and migration, and we found fucoidan to induce proliferation in 92.1 and OMM2.5 cells. Fucoidan has been shown to inhibit migration e.g., in colon, lung or bladder cancer cells [6,37,38]. Again, the effect is cell-type dependent, and no general anti-migratory effect of fucoidan could be shown here.

When we tested the ability of fucoidan to reduce the availability of VEGF, no reduction of VEGF could be seen at a concentration of 100 μg/mL. This is in contrast to our findings in retinal pigment epithelial (RPE) cells [33], where we could find a significant reduction of detectable VEGF at this concentration using the same batch of fucoidan A similar reductive effect of VEGF expression by fucoidan has been shown for breast cancer cells [9]. Therefore, the effect of fucoidan is not only determined by the molecular structure of the fucoidan [5], but also by the target cells. As higher concentrations of fucoidan did reduce VEGF in OMM1 cells, it is possible that the concentrations chosen in this experiment were too low to exert an effect. However, even in higher concentrations, the effect was cell type dependent, as OMM2.3 did not show any reduction of VEGF in any of the fucoidan concentrations tested. The pathways of fucoidan-mediated VEGF reduction have not been elucidated to date, but it has been shown that fucoidan can inhibit the activation of VEGFR-2 by preventing the binding of VEGF165 to its receptor [39]. We have previously shown that VEGF is autoregulated via the VEGFR-2 in RPE cells [40], and so we hypothesized that the downregulation of VEGF was mediated by interfering with the autoregulatory pathway. The cell dependent effect of fucoidan concerning VEGF in the UM cells may therefore be related to the presence of an autoregulatory pathway of VEGF expression in the tested melanoma cells.

In addition, in our angiogenesis assay, fucoidan induced the outgrowth of tubular structures, both in length and area, in 92.1 cells. Even though the general interaction between 92.1 and endothelial cells were low, this result may indicate that fucoidan may facilitate angiogenesis primary UM, which would not be desirable in patient treatment. Again, this cannot simply be explained by the molecular structure of this particular fucoidan, as we have shown before that this exact fucoidan reduced angiogenic structures in RPE-endothelial cells co-cultures [33].

Fucoidan displayed a significant protective effect against H_2O_2-induced cell death in all tested cell lines. Fucoidan has been reported to protect cells against oxidative stress [41,42]; however, to the best of our knowledge, this has not been shown in cancer cells before. Indeed, fucoidan, when given in addition with a chemotherapeutic, has been shown to increase oxidative stress in breast cancer cell [43]. Antioxidants may enhance tumor progression [20] and oxidative stress may protect from metastasis [21], so the protection of cancer cells against oxidative stress by fucoidan has to be taken into consideration when discussing fucoidan-derived drugs as possible new cancer agents [14]. Our data showed that the protective effects of fucoidan are not mediated via a change in the Bcl-2/Bax expression, or via the ERK1/2 or Akt pathway. Further research needs to be conducted in order to decipher the protective pathways of these compounds.

Fucoidan is also under investigation to be used in combination with other chemotherapeutic drugs in order to enhance their efficacy, as seen in e.g., melanoma [44] or breast cancer cells [43], where pro-apoptotic or anti-proliferative effects of the chemotherapeutics are enhanced by fucoidan. The results found in our study cannot be extrapolated towards combination treatments, however, the effect of fucoidan in combination treatments is also cell type dependent and may reduce the efficacy of the chemotherapeutic compound [45]. Moreover, it has been suggested that the apoptosis-enhancing effects of combination therapies combining fucoidan and chemotherapy is mediated by oxidative stress-enhancement by fucoidan [43], while our data show that fucoidan protects against oxidative stress. Therefore, our data cannot give a prediction about potential combination therapies in UM, and would strongly advise for caution in this area.

4. Conclusions

The data obtained in this study indicate that fucoidan is not suitable as a potential treatment for UM.

5. Material and Methods

5.1. Cell Culture of Melanoma Cells

Five established human UM cell lines were used. The cell lines 92.1 [46] and Mel270 [47] originated from primary UM, while all OMM cell lines are of metastatic origin; OMM2.5 and OMM2.3 from liver metastases [47] and OMM1 from a sub-cutaneous metastasis [48]. Cell cultures were maintained in RPMI (PAA Laboratories, Cölbe, Germany), supplemented with 10% fetal calf serum (FCS) (Linaris, Dossenheim, Germany) and 1% penicillin/streptomycin (PAA). Medium was exchanged three times a week and cells were passaged after reaching confluence.

5.2. Fucoidan

For the experiments, fucoidan from Sigma Aldrich (from Fucus vesiculosus, Sigma Aldrich, Steinheim, Germany; #F5631, [O28K3779; CAS 9072-19-9]) was used.

5.3. Proliferation

To determine the influence of fucoidan on proliferation, a defined number (200,000 cells) of the respective cell line was seeded on 12 well plates. Cells were stimulated with 100 µg/mL fucoidan for up to three days. In addition, for the cell lines OMM1 and OMM2.3, a dose-response curve after 24 h of incubation was determined, investigating 1 µg/mL, 10 µg/mL, 100 µg/mL and 1 mg/mL fucoidan. After the indicated period of time, a WST-assay was conducted.

5.4. WST-Assay

Treated cells as described above were treated with WST-1 reagent (Hoffmann-La Roche, Basel, Switzerland) for 4 h at 37°. The cells were rocked on a shaker for 2 min, the supernatant was collected, and measured at 450 nm.

5.5. Scratch Assay

The scratch assay was conducted as previously described with modifications [33]. In brief, the respective cell line was seeded in a 12-well-plate. Two wounds were scratched in the confluent cell layer with a pipette tip and the cells were washed with PBS to remove detached cells. Microscopic bright field pictures of three spots were taken (AxioCam, Zeiss, Jena, Germany). Fucoidan (100 µg/mL) was added to the wells. After 90% wound closure of the control, another picture was taken. To analyse the wound healing capability of the cells, application was conducted in duplicates and three pictures per well were taken. The gap size of the wound was measured with AxioVision Rel.4.8. (Zeiss, Jena, Germany), and the percentage of coverage of the wound was evaluated. Complete coverage was defined as 100%.

5.6. VEGF-ELISA

The supernatant of cell cultures was collected after 100 µM fucoidan incubation for up to three days. In addition, for the cell lines OMM1 and OMM2.3, a dose-response curve after 24 h of incubation was determined, investigating 1 µg/mL, 10 µg/mL, 100 µg/mL and 1 mg/mL fucoidan. VEGF-content was measured by VEGF-ELISA (R&D Systems, Wiesbaden, Germany), following the manufacturer's instructions. The range of detection of the ELISA was between 15 pg/mL and 1046 pg/mL. The amount of VEGF secreted was normalized to cell number. Cell number was assessed with a trypan blue exclusion assay.

5.7. Angiogenesis Assay

Angiogenesis was evaluated in a direct co-culture system of UM cells and outgrowth endothelial cells.

The isolation of outgrowth endothelial cells from peripheral blood was conducted as described previously [49,50]. In brief, these cells were isolated from buffy coats by isolation of blood mononuclear cells. Mononuclear cells were seeded onto collagen coated 24-well plates in a density of 5×10^6 cells/well in EGM-2 (Lonza, Basel, Switzerland) with full supplements from the kit, 5% FCS, and 1% penicillin/streptomycin. After one week, adherent cells were collected by trypsin and reseeded on collagen coated 24-well plates in a density of 0.6×10^6 cells/well. After 2–3 weeks, colonies of endothelial cells (OEC) were harvested and further expanded over several passages using EGM-2 in a splitting ratio of 1:2.

Co-culture assays were performed for one primary (92.1) and one metastatic (OMM2.3) melanoma cell line. For co-cultures 100,000 cells/cm^2 were seeded into fibronectin coated thermanox coverslips in 24 well plates in their respective cell culture medium. On the next day outgrowth endothelial cells (OEC) were added to the cultures in a density of 100,000 cells/cm^2 to the respective uveal melanoma cell line and co-cultures were further maintained for seven days in EGM-2 treated with 100 µg/mL fucoidan, respectively, or left untreated in control groups. After seven days, co-cultures were fixed with 4% paraformaldehyde and outgrowth endothelial cells were immunostained for the endothelial marker CD31. All cells are counterstained by Hoechst and pictures were taken with a confocal laser scanning microscope (Zeiss LSM 510 Meta, Jena, Germany). Angiogenesis was evaluated in comparison to untreated controls. For each group, at least three pictures were taken from two technical replicates. These experiments and the picture analysis were performed with endothelial cells from three different donors.

5.8. Image Analysis

The microscopic images were analyzed using the image processing program ImageJ Vers. 1.47 and GIMP 2.8. The analysis of angiogenic structures was conducted as previously described [51]. In brief, tube-like structures were extracted from the background by automatic segmentation after background correction. The binaries of the tube-like structures were further processed, including a final manual correction. The resulting binaries were analyzed for the area and the length of tubular structures. Additionally, the total area of fluorescence was assessed after automatic segmentation.

5.9. Cytotoxicity

Cells were plated on 24-well plates. H_2O_2 (Sigma-Aldrich, Munich, Germany) was applied in order to induce oxidative stress mediated cytotoxicity. We have previously shown that uveal melanoma cell lines show a cell-line specific susceptibility to oxidative stress [32]. Cytotoxicity was induced by applying H_2O_2 in the respective concentration (92.1:250 µM, Mel270 and OMM1: 500 µM, OMM2.3 and OMM2.5:1000 µM). In order to evaluate a potential protective effect of fucoidan, confluent cells were treated 30 min prior to oxidative insult with 100 µg/mL fucoidan. Cell viability was assessed after 24 h of stimulation with a WST assay.

5.10. Whole Cell Lysate

After treatment of cells as indicated, whole cell lysates were prepared in an NP-40 buffer as described previously [33]. In brief, cells were washed with PBS and NP-40 buffer (1% Nonidet® P40 Substitute, 150 mM NaCl, 50 mM Tris, pH 8.0) was added. The lysates were kept on ice for at least 30 min. Lysates were centrifuged at 13,000 rpm for 15 min and the supernatant harvested. The protein concentration of the supernatant was determined by a BioRad protein assay (BioRad, München, Germany) with bovine serum albumin (Fluka, Buchs, Switzerland) used as standard.

5.11. Western Blot

Western blot was conducted as described previously with modifications [52]. In brief, proteins were separated in an SDS-PAGE, using 12% acrylamide gels. Gels were blotted on PVDF-membranes (Carl Roth GmbH, Karlsruhe, Germany) and then blocked in 4% skim milk in Tris buffered saline with 0.1% Tween for 1 h at room temperature. The blot was treated with the first antibodies, beta-actin (#4967, 1:1000), Akt (#9272, 1:1000), ERK1/2 (#9102, 1:1000), p-ERK1/2 (#9101, 1:1000) (all Cell-Signaling Technologies, CST, Denver, CO, USA; all rabbit), Bax (sc-20067, 1:1000) or Bcl-2 (sc-509, 1:1000) (all Santa Cruz, Heidelberg, Germany, all mouse), respectively, in 2% skim milk in Tris buffered saline with 0.1% Tween overnight at 4 °C. After washing the blot, it was incubated with appropriate secondary antibody (anti-rabbit (#7074) or anti-mouse (#7076) IgG, HRP-linked antibody (all Cell-Signaling)) in 2% skim milk in Tris-buffered saline with 0.1% Tween (Merck, Darmstadt, Germany). Following the final wash, the blot was incubated with Immobilon chemiluminscence reagent (Merck), and the signal was detected with MF-ChemiBis 1.6 (Biostep, Jahnsdorf, Germany). The density of the bands was evaluated using Total lab software (Biostep) and the signal was normalized for ß-actin.

5.12. Statistics

Statistical analysis was performed with MS-Excel. Means ± standard deviation (sd) was calculated for at least three independent sets of experiments. Significant differences between means were calculated by *t*-test. A *p*-value of 0.05 or less was considered significant.

Acknowledgments: This study was supported by the Werner and Klara Kreitz-Foundation. We would like to thank Serap Luick and Andrea Hethke for their excellent technical assistance.

Author Contributions: A.K., M.D., S.F., and J.R. conceived and designed the experiments. S.E.C. and J.R. contributed materials. M.D., A.-M.K., F.W., E.R. and H.S. performed the experiments. M.D., A.-M.K., S.F., H.S., S.E.C., E.R. and A.K. analyzed the data. A.K. wrote the paper, which was reviewed by all authors.

Mar. Drugs **2017**, *15*, 193

Conflicts of Interest: The authors declare no conflict of interest.

References

1. Shields, J.A.; Shields, C.L. Management of posterior uveal melanoma: Past, present, and future: The 2014 Charles L. Schepens lecture. *Ophthalmology* **2015**, *122*, 414–428. [CrossRef] [PubMed]
2. Shields, C.L.; Kels, J.G.; Shields, J.A. Melanoma of the eye: Revealing hidden secrets, one at a time. *Clin. Dermatol.* **2015**, *33*, 183–196. [CrossRef] [PubMed]
3. Seregard, S.; Pelayes, D.E.; Singh, A.D. Radiation therapy: Uveal tumors. *Dev. Ophthalmol.* **2013**, *52*, 36–57. [PubMed]
4. Spagnolo, F.; Caltabiano, G.; Queirolo, P. Uveal melanoma. *Cancer Treat. Rev.* **2012**, *38*, 549–553. [CrossRef] [PubMed]
5. Wu, L.; Sun, J.; Su, X.; Yu, Q.; Yu, Q.; Zhang, P. A review about the development of fucoidan in antitumor activity: Progress and challenges. *Carbohydr. Polym.* **2016**, *154*, 96–111. [CrossRef] [PubMed]
6. Han, Y.S.; Lee, J.H.; Lee, S.H. Antitumor Effects of Fucoidan on Human Colon Cancer Cells via Activation of Akt Signaling. *Biomol. Ther.* **2015**, *23*, 225–232. [CrossRef] [PubMed]
7. Kawaguchi, T.; Hayakawa, M.; Koga, H.; Torimura, T. Effects of fucoidan on proliferation, AMP-activated protein kinase, and downstream metabolism- and cell cycle-associated molecules in poorly differentiated human hepatoma HLF cells. *Int. J. Oncol.* **2015**, *46*, 2216–2222. [CrossRef] [PubMed]
8. Park, H.Y.; Kim, G.Y.; Moon, S.K.; Kim, W.J.; Yoo, Y.H.; Choi, Y.H. Fucoidan inhibits the proliferation of human urinary bladder cancer T24 cells by blocking cell cycle progression and inducing apoptosis. *Molecules* **2014**, *19*, 5981–5998. [CrossRef] [PubMed]
9. Xue, M.; Ge, Y.; Zhang, J.; Wang, Q.; Hou, L.; Liu, Y.; Sun, L.; Li, Q. Anticancer properties and mechanisms of fucoidan on mouse breast cancer in vitro and in vivo. *PLoS ONE* **2012**, *7*, e43483. [CrossRef] [PubMed]
10. Ale, M.T.; Maruyama, H.; Tamauchi, H.; Mikkelsen, J.D.; Meyer, A.S. Fucoidan from *Sargassum* sp. and *Fucus vesiculosus* reduces cell viability of lung carcinoma and melanoma cells in vitro and activates natural killer cells in mice in vivo. *Int. J. Biol. Macromol.* **2011**, *49*, 331–336. [CrossRef] [PubMed]
11. Boo, H.J.; Hong, J.Y.; Kim, S.C.; Kang, J.I.; Kim, M.K.; Kim, E.J.; Hyun, J.W.; Koh, Y.S.; Yoo, E.S.; Kwon, J.M.; et al. The anticancer effect of fucoidan in PC-3 prostate cancer cells. *Mar. Drugs* **2013**, *11*, 2982–2999. [CrossRef] [PubMed]
12. Wang, W.; Chen, H.; Zhang, L.; Qin, Y.; Cong, Q.; Wang, P.; Ding, K. A fucoidan from *Nemacystus decipiens* disrupts angiogenesis through targeting bone morphogenetic protein 4. *Carbohydr. Polym.* **2016**, *144*, 305–314. [CrossRef] [PubMed]
13. Liu, F.; Wang, J.; Chang, A.K.; Liu, B.; Yang, L.; Li, Q.; Wang, P.; Zou, X. Fucoidan extract derived from *Undaria pinnatifida* inhibits angiogenesis by human umbilical vein endothelial cells. *Phytomedicine* **2012**, *19*, 797–803. [CrossRef] [PubMed]
14. Atashrazm, F.; Lowenthal, R.M.; Woods, G.M.; Holloway, A.F.; Dickinson, J.L. Fucoidan and cancer: A multifunctional molecule with anti-tumor potential. *Mar. Drugs* **2015**, *13*, 2327–2346. [CrossRef] [PubMed]
15. Missotten, G.S.; Notting, I.C.; Schlingemann, R.O.; Zijlmans, H.J.; Lau, C.; Eilers, P.H.; Keunen, J.E.; Jager, M.J. Vascular endothelial growth factor a in eyes with uveal melanoma. *Arch. Ophthalmol.* **2006**, *124*, 1428–1434. [CrossRef] [PubMed]
16. Boyd, S.R.; Tan, D.; Bunce, C.; Gittos, A.; Neale, M.H.; Hungerford, J.L.; Charnock-Jones, S.; Cree, I.A. Vascular endothelial growth factor is elevated in ocular fluids of eyes harbouring uveal melanoma: Identification of a potential therapeutic window. *Br. J. Ophthalmol.* **2002**, *86*, 448–452. [CrossRef] [PubMed]
17. El Filali, M.; Missotten, G.S.; Maat, W.; Ly, L.V.; Luyten, G.P.; van der Velden, P.A.; Jager, M.J. Regulation of VEGF-A in uveal melanoma. *Investig. Ophthalmol. Vis. Sci.* **2010**, *51*, 2329–2337. [CrossRef] [PubMed]
18. Yang, M.; Kuang, X.; Pan, Y.; Tan, M.; Lu, B.; Lu, J.; Cheng, Q.; Li, J. Clinicopathological characteristics of vascular endothelial growth factor expression in uveal melanoma: A meta-analysis. *Mol. Clin. Oncol.* **2014**, *2*, 363–368. [PubMed]
19. Huang, T.H.; Chiu, Y.H.; Chan, Y.L.; Chiu, Y.H.; Wang, H.; Huang, K.C.; Li, T.L.; Hsu, K.H.; Wu, C.J. Prophylactic administration of fucoidan represses cancer metastasis by inhibiting vascular endothelial growth factor (VEGF) and matrix metalloproteinases (MMPs) in Lewis tumor-bearing mice. *Mar. Drugs* **2015**, *13*, 1882–1900. [CrossRef] [PubMed]

20. Harris, I.S.; Treloar, A.E.; Inoue, S.; Sasaki, M.; Gorrini, C.; Lee, K.C.; Yung, K.Y.; Brenner, D.; Knobbe-Thomsen, C.B.; Cox, M.A.; et al. Glutathione and thioredoxin antioxidant pathways synergize to drive cancer initiation and progression. *Cancer Cell* **2015**, *27*, 211–222. [CrossRef] [PubMed]

21. Piskounova, E.; Agathocleous, M.; Murphy, M.M.; Hu, Z.; Huddlestun, S.E.; Zhao, Z.; Leitch, A.M.; Johnson, T.M.; DeBerardinis, R.J.; Morrison, S.J. Oxidative stress inhibits distant metastasis by human melanoma cells. *Nature* **2015**, *527*, 186–191. [CrossRef] [PubMed]

22. Gorrini, C.; Harris, I.S.; Mak, T.W. Modulation of oxidative stress as an anticancer strategy. *Nat. Rev. Drug Discov.* **2013**, *12*, 931–947. [CrossRef] [PubMed]

23. Leach, J.K.; Van Tuyle, G.; Lin, P.S.; Schmidt-Ullrich, R.; Mikkelsen, R.B. Ionizing radiation-induced, mitochondria-dependent generation of reactive oxygen/nitrogen. *Cancer Res.* **2001**, *61*, 3894–3901. [PubMed]

24. Hyun, J.H.; Kim, S.C.; Kang, J.I.; Kim, M.K.; Boo, H.J.; Kwon, J.M.; Koh, Y.S.; Hyun, J.W.; Park, D.B.; Yoo, E.S.; et al. Apoptosis inducing activity of fucoidan in HCT-15 colon carcinoma cells. *Biol. Pharm. Bull.* **2009**, *32*, 1760–1764. [CrossRef] [PubMed]

25. Park, H.S.; Hwang, H.J.; Kim, G.Y.; Cha, H.J.; Kim, W.J.; Kim, N.D.; Yoo, Y.H.; Choi, Y.H. Induction of apoptosis by fucoidan in human leukemia U937 cells through activation of p38 MAPK and modulation of Bcl-2 family. *Mar. Drugs* **2013**, *11*, 2347–2364. [CrossRef] [PubMed]

26. Lefèvre, G.; Babchia, N.; Calipel, A.; Mouriaux, F.; Faussat, A.M.; Mrzyk, S.; Mascarelli, F. Activation of the FGF2/FGFR1 autocrine loop for cell proliferation and survival in uveal melanoma cells. *Investig. Ophthalmol. Vis. Sci.* **2009**, *50*, 1047–1057. [CrossRef] [PubMed]

27. Babchia, N.; Calipel, A.; Mouriaux, F.; Faussat, A.M.; Mascarelli, F. The PI3K/Akt and mTOR/P70S6K signaling pathways in human uveal melanoma cells: Interaction with B-Raf/ERK. *Investig. Ophthalmol. Vis. Sci.* **2010**, *51*, 421–429. [CrossRef] [PubMed]

28. Samadi, A.K.; Cohen, S.M.; Mukerji, R.; Chaguturu, V.; Zhang, X.; Timmermann, B.N.; Cohen, M.S.; Person, E.A. Natural withanolide withaferin A induces apoptosis in uveal melanoma cells by suppression of Akt and c-MET activation. *Tumour Biol.* **2012**, *33*, 1179–1189. [CrossRef] [PubMed]

29. Ho, A.L.; Musi, E.; Ambrosini, G.; Nair, J.S.; Deraje Vasudeva, S.; de Stanchina, E.; Schwartz, G.K. Impact of combined mTOR and MEK inhibition in uveal melanoma is driven by tumor genotype. *PLoS ONE* **2012**, *7*, e40439. [CrossRef] [PubMed]

30. Wang, J.; Jia, R.; Zhang, Y.; Xu, X.; Song, X.; Zhou, Y.; Zhang, H.; Ge, S.; Fan, X. The role of Bax and Bcl-2 in gemcitabine-mediated cytotoxicity in uveal melanoma cells. *Tumour Biol.* **2014**, *35*, 1169–1175. [CrossRef] [PubMed]

31. Sulkowska, M.; Famulski, W.; Bakunowicz-Lazarczyk, A.; Chyczewski, L.; Sulkowski, S. Bcl-2 expression in primary uveal melanoma. *Tumori* **2001**, *87*, 54–57. [PubMed]

32. Dithmer, M.; Kirsch, A.M.; Gräfenstein, L.; Wang, F.; Schmidt, H.; Coupland, S.E.; Fuchs, S.; Roider, J.; Klettner, A. Uveale Melanomzellen unter oxidativen Stress—Einfluss von VEGF und VEGF-Inhibitoren. *Klin. Monatsbl. Augenhkd.* **2017**. [CrossRef]

33. Dithmer, M.; Fuchs, S.; Shi, Y.; Schmidt, H.; Richert, E.; Roider, J.; Klettner, A. Fucoidan reduces secretion and expression of vascular endothelial growth factor in the retinal pigment epithelium and reduces angiogenesis in vitro. *PLoS ONE* **2014**, *9*, e89150. [CrossRef] [PubMed]

34. Riou, D.; Colliec-Jouault, S.; Pinczon du Sel, D.; Bosch, S.; Siavoshian, S.; Le Bert, V.; Tomasoni, C.; Sinquin, C.; Durand, P.; Roussakis, C. Antitumor and antiproliferative effects of a fucan extracted from ascophyllum nodosum against a non-small-cell bronchopulmonary carcinoma line. *Anticancer Res.* **1996**, *16*, 1213–1218. [PubMed]

35. Ale, M.T.; Maruyama, H.; Tamauchi, H.; Mikkelsen, J.D.; Meyer, A.S. Fucose-containing sulfated polysaccharides from brown seaweeds inhibit proliferation of melanoma cells and induce apoptosis by activation of caspase-3 in vitro. *Mar. Drugs* **2011**, *9*, 2605–2621. [CrossRef] [PubMed]

36. Yang, G.; Zhang, Q.; Kong, Y.; Xie, B.; Gao, M.; Tao, Y.; Xu, H.; Zhan, F.; Dai, B.; Shi, J.; et al. Antitumor activity of fucoidan against diffuse large B cell lymphoma in vitro and in vivo. *Acta Biochim. Biophys. Sin.* **2015**, *47*, 925–931. [CrossRef] [PubMed]

37. Cho, T.M.; Kim, W.J.; Moon, S.K. AKT signaling is involved in fucoidan-induced inhibition of growth and migration of human bladder cancer cells. *Food Chem. Toxicol.* **2014**, *64*, 344–352. [CrossRef] [PubMed]

38. Lee, H.; Kim, J.S.; Kim, E. Fucoidan from Seaweed Fucus vesiculosus Inhibits Migration and Invasion of Human Lung Cancer Cell via PI3K-Akt-mTOR Pathways. *PLoS ONE* **2012**, *7*, e50624. [CrossRef] [PubMed]

39. Koyanagi, S.; Tanigawa, N.; Nakagawa, H.; Soeda, S.; Shimeno, H. Oversulfation of fucoidan enhances its anti-angiogenic and antitumor activities. *Biochem. Pharmacol.* **2003**, *65*, 173–179. [CrossRef]

40. Klettner, A.; Westhues, D.; Lassen, J.; Bartsch, S.; Roider, J. Regulation of constitutive vascular endothelial growth factor secretion in retinal pigment epithelium/choroid organ cultures: P38, nuclear factor κB, and the vascular endothelial growth factor receptor-2/phosphatidylinositol 3 kinase pathway. *Mol. Vis.* **2013**, *19*, 281–291. [PubMed]

41. Han, Y.S.; Lee, J.H.; Jung, J.S.; Noh, H.; Baek, M.J.; Ryu, J.M.; Yoon, Y.M.; Han, H.J.; Lee, S.H. Fucoidan protects mesenchymal stem cells against oxidative stress and enhances vascular regeneration in a murine hindlimb ischemia model. *Int. J. Cardiol.* **2015**, *198*, 187–195. [CrossRef] [PubMed]

42. Li, X.; Zhao, H.; Wang, Q.; Liang, H.; Jiang, X. Fucoidan protects ARPE-19 cells from oxidative stress via normalization of reactive oxygen species generation through the Ca^{2+}-dependent ERK signaling pathway. *Mol. Med. Rep.* **2015**, *11*, 3746–3752. [CrossRef] [PubMed]

43. Zhang, Z.; Teruya, K.; Yoshida, T.; Eto, H.; Shirahata, S. Fucoidan extract enhances the anti-cancer activity of chemotherapeutic agents in MDA-MB-231 and MCF-7 breast cancer cells. *Mar. Drugs* **2013**, *11*, 81–98. [CrossRef] [PubMed]

44. Thakur, V.; Lu, J.; Roscilli, G.; Aurisicchio, L.; Cappelletti, M.; Pavoni, E.; White, W.L.; Bedogni, B. The natural compound fucoidan from New Zealand Undaria pinnatifida synergizes with the ERBB inhibitor lapatinib enhancing melanoma growth inhibition. *Oncotarget* **2017**, *8*, 17887–17896. [CrossRef] [PubMed]

45. Oh, B.; Kim, J.; Lu, W.; Rosenthal, D. Anticancer Effect of Fucoidan in Combination with Tyrosine Kinase Inhibitor Lapatinib. *Evid.-Based Complement. Altern. Med.* **2014**, *2014*, 865375. [CrossRef] [PubMed]

46. De Waard-Siebinga, I.; Blom, D.J.; Griffioen, M.; Schrier, PI.; Hoogendoorn, E.; Beverstock, G.; Danen, E.H.; Jager, M.J. Establishment and characterization of an uveal-melanoma cell line. *Int. J. Cancer* **1995**, *62*, 155–161. [CrossRef] [PubMed]

47. Verbik, D.J.; Murray, T.G.; Tran, J.M.; Ksander, B.R. Melanomas that develop within the eye inhibit lymphocyte proliferation. *Int. J. Cancer* **1997**, *73*, 470–478. [CrossRef]

48. Luyten, G.P.; Naus, N.C.; Mooy, C.M.; Hagemeijer, A.; Kan-Mitchell, J.; Van Drunen, E.; Vuzevski, V.; De Jong, P.T.; Luider, T.M. Establishment and characterization of primary and metastatic uveal melanoma cell lines. *Int. J. Cancer* **1996**, *66*, 380–387. [CrossRef]

49. Fuchs, S.; Motta, A.; Migliaresi, C.; Kirkpatrick, C.J. Outgrowth endothelial cells isolated and expanded from human peripheral blood progenitor cells as a potential source of autologous cells for endothelialization of silk fibroin biomaterials. *Biomaterials* **2006**, *27*, 5399–5408. [CrossRef] [PubMed]

50. Fuchs, S.; Hofmann, A.; Kirkpartrick, C. Microvessel-like structures from outgrowth endothelial cells from human peripheral blood in 2-dimensional and 3-dimensional co-cultures with osteoblastic lineage cells. *Tissue Eng.* **2007**, *13*, 2577–2588. [CrossRef] [PubMed]

51. Fuchs, S.; Jiang, X.; Schmidt, H.; Dohle, E.; Ghanaati, S.; Orth, C.; Hofmann, A.; Motta, A.; Migliaresi, C.; Kirkpatrick, C.J. Dynamic processes involved in the pre-vascularization of silk fibroin constructs for bone regeneration using outgrowth endothelial cells. *Biomaterials* **2009**, *30*, 1329–1338. [CrossRef] [PubMed]

52. Faby, H.; Hillenkamp, J.; Roider, J.; Klettner, A. Hyperthermia-induced upregulation of vascular endothelial growth factor in retinal pigment epithelial cells is regulated by mitogen-activated protein kinases. *Graefes Arch. Clin. Exp. Ophthalmol.* **2014**, *252*, 1737–1745. [CrossRef] [PubMed]

marine drugs

MDPI

Article

Cembrene Diterpenoids with Ether Linkages from *Sarcophyton ehrenbergi*: An Anti-Proliferation and Molecular-Docking Assessment

Mohamed-Elamir F. Hegazy [1], Abdelsamed I. Elshamy [2], Tarik A. Mohamed [1], Ahmed R. Hamed [1,3], Mahmoud A. A. Ibrahim [4], Shinji Ohta [5] and Paul W. Paré [6,*]

[1] Phytochemistry Department, National Research Centre, 33 El-Bohouth St., Dokki, Giza 12622, Egypt; elamir77@live.com (M.-E.F.H.); tarik.nrc83@yahoo.com (T.A.M.); n1ragab2004@yahoo.com (A.R.H.)
[2] Natural Compounds Chemistry Department, National Research Centre, 33 El-Bohouth St., Dokki, Giza 12622, Egypt; elshamynrc@yahoo.com
[3] Biology Unit, Central Laboratory for Pharmaceutical and Drug Industries Research Division, National Research Centre, 33 El-Bohouth St., Dokki, Giza 12622, Egypt
[4] Computational Chemistry Laboratory, Chemistry Department, Faculty of Science, Minia University, Minia 61519, Egypt; m.ibrahim@compchem.net
[5] Graduate School of Biosphere Science, Hiroshima University, 1-7-1 Kagamiyama, Higashi-Hiroshima 739-8521, Japan; ohta@hiroshima-u.ac.jp
[6] Department of Chemistry and Biochemistry, Texas Tech University, Lubbock, TX 79409, USA
* Correspondence: paul.pare@ttu.edu; Tel.: +20-122-007-3557; Fax: +20-233-370-931

Received: 3 May 2017; Accepted: 14 June 2017; Published: 21 June 2017

Abstract: Three new cembrene diterpenoids, sarcoehrenbergilid A–C (**1**–**3**), along with four known diterpenoids, sarcophine (**4**), (+)-7α,8β-dihydroxydeepoxysarcophine (**5**), sinulolide A (**6**), and sinulolide B (**7**), and one steroid, sardisterol (**8**), were isolated and characterized from a solvent extract of the Red Sea soft coral *Sarcophyton ehrenbergi*. Chemical structures were elucidated by NMR and MS analyses with absolute stereochemistry determined by X-ray analysis. Since these isolated cembrene diterpenes contained 10 or more carbons in a large flexible ring, conformer stabilities were examined based on density functional theory calculations. Anti-proliferative activities for **1**–**8** were evaluated against three human tumor cell lines of different origins including the: lung (A549), colon (Caco-2), and liver (HepG2). Sardisterol (**8**) was the most potent of the metabolites isolated with an IC$_{50}$ of 27.3 μM against the A549 cell line. Since an elevated human-cancer occurrence is associated with an aberrant receptor function for the epidermal growth factor receptor (EGFR), molecular docking studies were used to examine preferential metabolite interactions/binding and probe the mode-of-action for metabolite-anti tumor activity.

Keywords: *Sarcophyton ehrenbergi*; soft coral; terpenes; cembranoids; cytotoxic activity; molecular docking

---

1. Introduction

Cembrane diterpenoids are a large and structurally diverse group of natural products isolated from both terrestrial and marine organisms [1]. The 14-membered ring structure is biosynthetically formed from the cyclization of the geranylgeraniol precursor between carbons 1 and 14. The cembranoid diterpene, sarcophytol A, first isolated from the Okinawan soft coral *Sarcophyton glaucum* and found to exhibit strong inhibitory activity against tumor promoters [2], led to the subsequent isolation of hundreds of cytotoxic cembranoids from plant and marine sources [3]. *Sarcophyton* soft coral species are characterized by the production of cembrene-type diterpenoids [4–10] and these cyclic diterpenes usually exhibit cyclic ether, lactone, or furane moieties around the cembrane framework [11,12]. From a biomedical perspective, cembranoid diterpenes exhibit a diverse range of biological protection against tumors, inflammation, and fish toxins (ichthyotoxic), as well as microbial and/or viral infections [4,5,13]. With cancer occurrence and mortality associated with cancer increasing in the U.S. and around the world, the exploration of cytotoxic agents for improved cancer treatment via chemotherapy is a global priority. In fact, the World Health Organization (WHO) estimates that malignant neoplasms are ranked as the second leading cause of death globally. In 2012, 14.1 million newly-diagnosed cancer cases were reported, with 8.2 million deaths directly associated with cancer; these incidence and mortality numbers are estimated to increase by ca. 150% by 2030 [14,15].

In a continuing effort to characterize soft coral metabolites from the Red Sea with biological activity [6–8], herein is reported three new cembrene diterpenoids, as well as known diterpenoids and a polyoxygenated steroid isolated from *Sarcophyton ehrenbergi*. In this study, the anti-proliferative potential of the isolated compounds against three human tumor cell lines were evaluated. To probe the mode-of-action, molecular docking studies were performed with the epidermal growth factor receptor (EGFR), a large family of transmembrane receptors that normally regulate key events associated with cell growth, differentiation, and migration. An aberrant receptor function has been linked to elevated cancer occurrence.

2. Results and Discussion

2.1. Identification and Structure Elucidation

As part of the continuing investigation for biologically active constituents from Egyptian Red Sea costal soft corals [6–8,16], reported here is the chromatographic fractionation and purification of a methylene chloride:methanol (1:1) extract from *S. ehrenbergi* (Figure 1).

Figure 1. Soft coral *Sarcophyton ehrenbergi* photographed in its native Red Sea habitat; the width of the species shown is ca. 20 cm.

Compound **1** was obtained as white crystals with an optical rotation of $[\alpha]_{25}^{D}$ −6.9 in $CHCl_3$. HRESIFTMS analysis showed a molecular ion peak at m/z 387.2142 [M + Na]$^+$ (calcd. 364.2250), corresponding to the molecular formula of $C_{21}H_{32}O_5$. The IR spectrum showed characteristic bands at 3450 cm^{-1} (OH) and 1754 cm^{-1} (CO). The ^{1}H NMR spectrum (Table 1) exhibited three oxygenated protons at δ_H 5.50 (d; J = 10.10 Hz); δ_H 3.24 (d; J = 6.90 Hz); and δ_H 3.37 (m). Only one olefinic proton at δ_H 4.94 d; J = 10.10 Hz was attributed to a tri-substituted double bond; four signals at δ_H 1.76 s, 1.86 s, 1.05 s, and 0.98 s were identified as methyls, in addition to one methyl of a methoxy group at δ_H 3.13 s. Twenty one carbon signals were observed in the ^{13}C NMR spectrum and classified by DEPT analysis as five methyls (including one methyl of the methoxy group at δ_C 49.0), six methylenes, four methines, and six quaternary carbons (including the carbonyl group of the lactone ring δ_C 175.9 (Table 1)). The spectrum also revealed the presence of four olefinic carbon signals at δ_C 122.3, 164.3, 119.4, and 141.6; three oxymethine carbons at δ_C 81.0, 78.5, and 79.0; and one oxygenated quaternary carbon of an epoxy at δ_C 78.3. The most oxygenated down-field carbon signal indicated the presence of an ether linkage that was functionality confirmed by HRESIFTMS. These spectroscopic data were consistent with a cembrene diterpenoid based on spectroscopic data reported for other *Sarcophyton* species [4,13] (Figure 2). Six degrees of unsaturation were deduced, suggesting a tricyclic skeleton. The correlation of the oxygenated proton at δ_H 5.50 (d; J = 10.10 Hz) with the olefinic signal at δ_H 4.94 (d, J = 10.10 Hz) in DQF-COSY, as well as with quaternary olefinic carbons at δ_C 141.6 and δ_C 164.3, allowed the assignments of H-2, H-3, C-4, and C-1 of a cembrene diterpenoid, respectively [9,10].

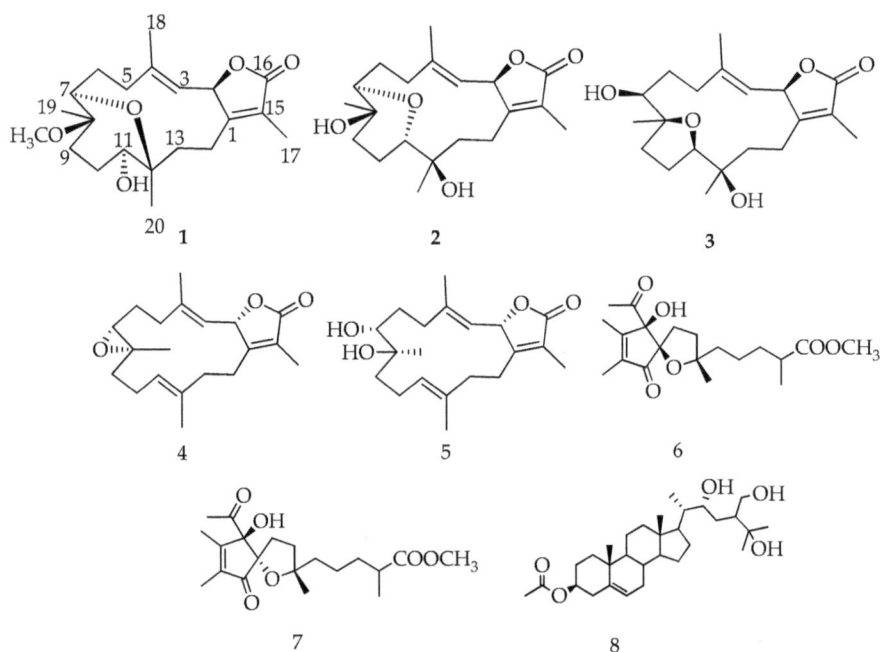

Figure 2. Structures of metabolites **1–8**.

The HMBC correlation of a methyl signal at δ_H 1.76 (s) with C-1 and a keto group at δ_C 175.9 allowed for the assignment of H-17 and C-16, respectively, and indicated the location of a lactone ring, including C-1/C-2. The observed HMBC correlation between H-3 and an olefinic methyl signal at δ_C 17.0 and a methylene signal at δ_C 40.9 allowed for the assignment of H-18 (δ_H 1.86, s) and H-5 [δ_H 2.07, t (J = 13.08)], respectively, which was confirmed by HMQC analysis. A doublet oxygenated methine signal at δ_H 3.24 (J = 6.90) correlated with a methyelene multiplet at δ_H 2.44/2.14 in DQF-COSY and C-5 in HMBC allowed for the assignment of H-7 and H$_2$-6, respectively. Additionally, HMBC correlations of the methyl singlet at δ_H 1.05 with H-7 and an oxygenated quaternary carbon atom at δ_C 80.0, as well as the methylene signal at δ_C 36.4, allowed for the assignment of H$_3$-19 (δ_C 17.1), C-8, and C-9, respectively. The oxygenated signal at δ_H 3.37 (m) was assigned to H-11 (δ_C 79.0) based on an HMBC correlation with C-9 and a methyl signal at δ_C 17.6 (C-20). Correlations were observed between δ_H 1.45 (m, H-13)/δ_H 1.95 (m, H-14) and C-20 in DQF-COSY and HMBC analyses, respectively (Figure 3). The location of a characteristic methoxy group signal at δ_H 3.13 (δ_C 49.0) was confirmed to be at C-8 via an HMBC correlation. The complete assignment of **1**, as well as the ether linkage between C-7/C-12 and the presence of the hydroxyl group at C-11, were established by NMR and HRESIFTMS data; structural confirmation including absolute configuration was established unambiguously using the anomalous scattering of Cu Kα radiation with the Flack parameter [17,18] being refined to 0.09 (3) (Figure 4).

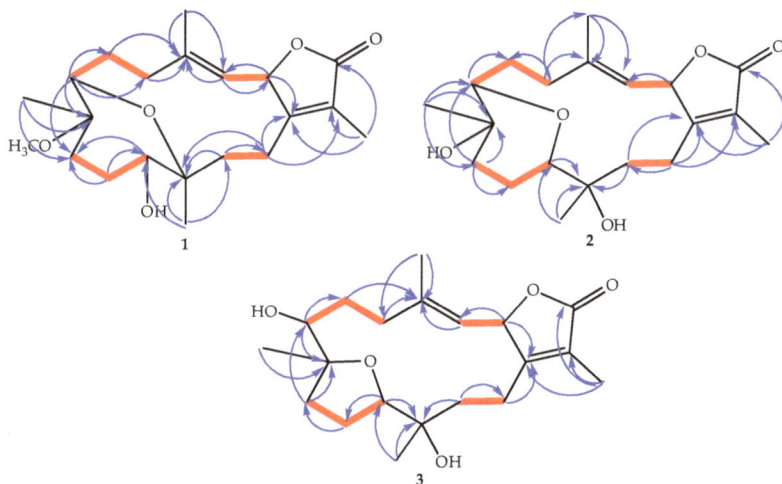

Figure 3. Selected ^{1}H-^{1}H COSY (–) and HMBC (➝) correlations of **1–3**.

The γ-lactone- (H-2) and olefinic-proton (H-3) vicinal coupling (10.10 Hz) established a *cis* configuration [8]. The four methyl groups exhibited NOSEY correlations with alpha protons consistent with the X-ray assignment of all methyl groups below the ring (e.g., CH$_3$-17 with H-14a, CH$_3$-18 with H-2, CH$_3$-19 with H-6a/H-10a, and CH$_3$-20 with H-10a) and absolute stereochemistry of 8*R* and 12*S* (Figure 5). NOSEY correlations between H-7 and H-5b, as well as H-11 and H-14b, were also consistent with 7*R* and 11*R* configurations. From this consistent x-ray and NMR data, **1** was assigned as 2*S*,16:7*R*,12*S*-diepoxy-11α-hydroxy-8β-methoxy-16-keto-cembra-1*E*,3*E*-diene (sarcoehrenbergilid A).

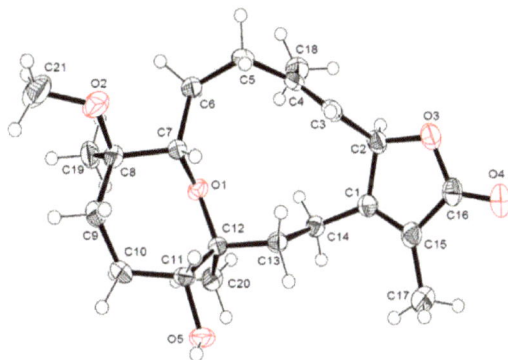

Figure 4. ORTEP depictions of cembrenoid **1** with oxygens (O1–O5) labeled in red.

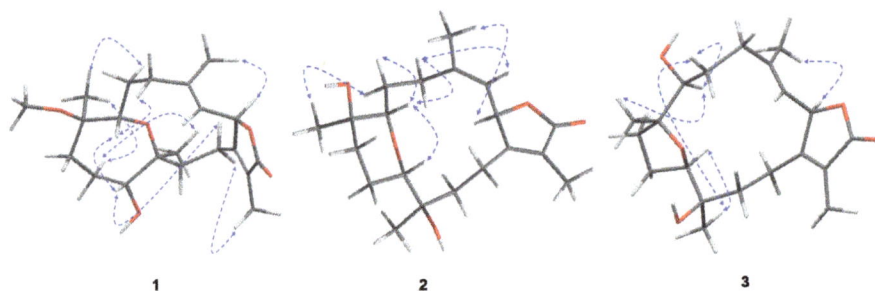

Figure 5. NOESY correlations for **1–3**.

Table 1. ^{1}H and ^{13}C NMR spectral data of **1–3** [a].

No.	1		2		3	
	δ_H	δ_C	δ_H	δ_C	δ_H	δ_C
1	——	164.3	——	164.6	——	164.4
2	5.50 d (10.1)	81.0	5.49 dd (10.1; 1.8)	81.7	5.51 br d (10.3)	79.6
3	4.94 d (10.1)	119.4	4.99 d (10.1)	120.0	4.87 br d (10.3)	117.8
4	——	141.6	——	142.6	——	146.0
5	2.07 t (13.1) 2.30 m	40.9	2.17 br t (13.1) 2.47 dd (10.1; 13.1)	38.8	2.09 br t (13.0) 2.41 br d (13.0)	40.3
6	2.14 dd (6.8), 1.44 m	27.6	1.66 m; 2.20 m	22.9	1.88 m; 1.58 m	24.9
7	3.24 d (6.9)	78.5	3.44 dd (11.6; 3.6)	69.7	3.01 br d (10.0)	86.8
8	——	80.0	——	73.6	——	69.4
9	1.53 m; 1.91 m	36.4	1.60 m; 2.40 m	36.6	1.88 m; 2.41 br d (13.0)	39.9
10	1.79 m; 2.15 m	28.5	1.61 m; 1.90 m	19.9	1.71 br d (10.9) 1.51 br t (10.9)	23.1
11	3.37 m	79.0	3.37 br d (11.9)	85.1	3.30 br d (10.9)	78.5
12	——	78.3	——	70.1	——	72.8
13	1.45 m; 2.37 m	34.6	1.59 m 1.78 dd (12.5; 3.7)	31.2	1.41 td (13.0; 5.8) 1.78 br t (13.0; 2.0)	35.0
14	1.95 br t (12.2) 2.31 td (12.2, 7.0)	20.8	2.00 br t (12.8) 2.58 td (12.8; 7.0)	21.1	2.17 br t (13.0); 2.67 m	21.1
15	——	122.3	——	122.3	——	122.5
16	——	175.9	——	176.0	——	175.6
17	1.76 s	8.8	1.83 s	8.7	1.82 s	8.7
18	1.86 s	17.0	1.92 br s	16.7	1.85 br s	16.1
19	1.05 s	17.1	1.22 s	22.1	1.15 s	20.0
20	0.98 s	17.6	1.13 s	25.3	1.16 s	23.8
21	3.13 s	49.0				

J values (Hz) are in parentheses; [a] Recorded in CDCl$_3$ and obtained at 600 and 150 MHz for ^{1}H and ^{13}C NMR, respectively.

Compound **2** was obtained as a white powder with an optical rotation of $[\alpha]_{25}^{D}$ −3.7 (*c* 0.0027, CHCl$_3$). HRESIFTMS analysis showed a molecular ion peak at *m*/*z* 373.1986 [(M + Na)$^+$] (calcd. 350.2093), implying six degrees of unsaturation. The IR spectrum exhibited characteristic bands at 3447 cm^{-1} (OH) and 1747 cm^{-1} (CO). ^{13}C NMR and DEPT spectral data (Table 1) showed 20 carbon resonances that distributed in the configuration of four methyls, six methylenes, four methines, and six quaternary carbons. Chemical shift data indicated the same cembrenoid backbone, containing diagnostic carbon signals associated with the lactone ring including a carbonyl signal C-16 (δ_C 175.6), three olefinic carbons at C-15, C-1, C-3, and C-4 (δ_C 122.5, 164.4, 117.8, and 146.0, respectively), and C-2 (δ_C 79.6). The spectra data closely matched a cemberene compound reported by Sawant et al. in 2004 [19], except for a large down field carbon signal difference at C-12 compared with the previously published structure tertiary carbon, which had a methyl substitution (δ_C 38.0). Compound **2** was proposed to contain a hydroxylated quaternary carbon at C-12 which would explain the downfield shift to δ_C72.8 and a 16 AMU addition compared to the previously published compound [20]. The addition of a hydroxyl group at C-12 was consistent with a H$_3$-20 downfield shift from δ_H 0.88 to δ_H 1.16 and a C-20 downfield shift from δ_C 17.2 to δ_C 23.8 without versus with a C-12 hydroxyl group. HMBC correlations of H$_3$-20 (δ_H 1.16 s) with C-12 (δ_C 72.8), C-11 (δ_C 78.5), and C-13 (δ_C 35.0) were also consistent with the hydroxylation of C-12 (Figure 3).

Similar to **1**, the γ-lactone- (H-2) and olefinic-proton (H-3) vicinal coupling (10 Hz) established a *cis* configuration [8]. Also similar to **1**, the NOSEY data for **2** showed a correlation between H-6b and H-7, indicating that the epoxy ring at C-7 is below the ring while H-6a correlates with CH$_3$-19, establishing that the relative stereochemistry for the methyl is above the ring (Figure 5). H-7, which is assumed to be in the beta position from the previous NOSEY correlation, also correlates with H-11, indicating that the other epoxide ring attachment is in an alpha configuration. Finally, a NOSEY correlation between H-10a and CH$_3$-20 indicates that the methyl group is in an alpha orientation. Thus, **2** was confirmed to be 2*S*,16: 7*R*,11*R*-diepoxy-8β,12β-dihydroxy-16-keto-cembra-1*E*,3*E*-diene (sarcoehrenbergilid B).

Compound **3** was obtained as a white powder with a negative optical rotation of $[\alpha]_{25}^D$ −6.6. HRESIFTMS analysis exhibited a molecular ion peak at *m/z* 373.1985 [(M + Na)$^+$] (calcd. 350.2093), corresponding to the molecular formula C$_{20}$H$_{30}$O$_5$ with six degrees of unsaturation. The IR spectrum showed characteristic bands at 3445 cm^{-1} (OH) and 1747 cm^{-1} (CO). Twenty carbon resonances were exhibited in the ^{13}C NMR and DEPT spectrum (Table 1); four methyls, six methylenes, five methines, and five quaternary carbons. The spectroscopic data of **3** are similar to a previously isolated diterpenoid from *S. trocheliophorum*, trocheliophorol [21], except for the presence of a hydroxyl unit at C-12 (δ_C 70.1) instead of an exomethylene. The location of the C-12 hydroxyl group was confirmed by HMBC correlations with a methyl singlet H$_3$-20 (δ_H 1.13 s); correlations were also observed between H$_3$-20 and δ_C 85.1 (C-11) and δ_C 31.2 (C-13) (Figure 3).

Similar to compounds **1** and **2**, a γ-lactone- (H-2) and olefinic-proton (H-3) vicinal coupling (10 Hz) established a *cis* configuration [8]. A NOESY correlation between H-6a with H-7 indicated that the C-7 hydroxyl group is in a beta orientation and a H-6a correlation with CH$_3$-19 indicated that the methyl at C-8 is in an alpha configuration (Figure 5). A NOSEY correlation was also observed between H-6a and CH$_3$-20, indicating that the hydroxyl at C-12 is in the beta orientation. Finally, a NOSEY correlation between CH$_3$-20 and H-11 indicated that the epoxide connection at C-11 is the same as C-8, both in alpha configurations. From the above spectral data, **3** was established as 2*S*,16: 8*S*,11*S*-diepoxy-7β,12α-dihydroxy-16-keto-cembra-1*E*,3*E*-diene (sarcoehrenbergilid C).

With the large flexible ring systems for compounds **1–3**, many conformers are possible. To predict the most stable form, molecular modeling calculations were performed to estimate the lowest energy state. Conformation ensembles were generated using a MMFF94 molecular mechanics force field within a 10 kJ/mol window, providing 101, 98, and 106 conforms for **1**, **2**, and **3**, respectively. Each generated conformer was subjected to energy-minimization at a B3LYP/6-31G* level of theory and thereafter, the corresponding free energy was calculated on the optimized structure. The Boltzmann population was estimated based on the calculated relative free energies for each conformer with respect to the lowest free energy conformer at 298 K (Table S1). The lowest and next three higher free energy conformers for **1–3** are shown (Figure 6). All conformations with populations higher than 1% are shown in Figures S22–S24. The optical rotation of all conformations was theoretically calculated and the results are summarized in Table S1. According to the calculated free energies and optical rotation, the lowest compound conform free energy is in strong agreement with the elucidated structures (Figure 6). For **2**, the structure is more stable than a previously reported conformer (Figure S23, **2f**) by −7.4 kJ/mol.

In addition to the three new metabolites (**1–3**), four known compounds, sarcophine (**4**), (+)-7α,8β-dihydroxydeepoxysarcophine (**5**) [6], sinulolide A (**6**), and B (**7**) [22], and one steroid, sardisterol (**8**) [23], were identified from the coral extract (Figure 2).

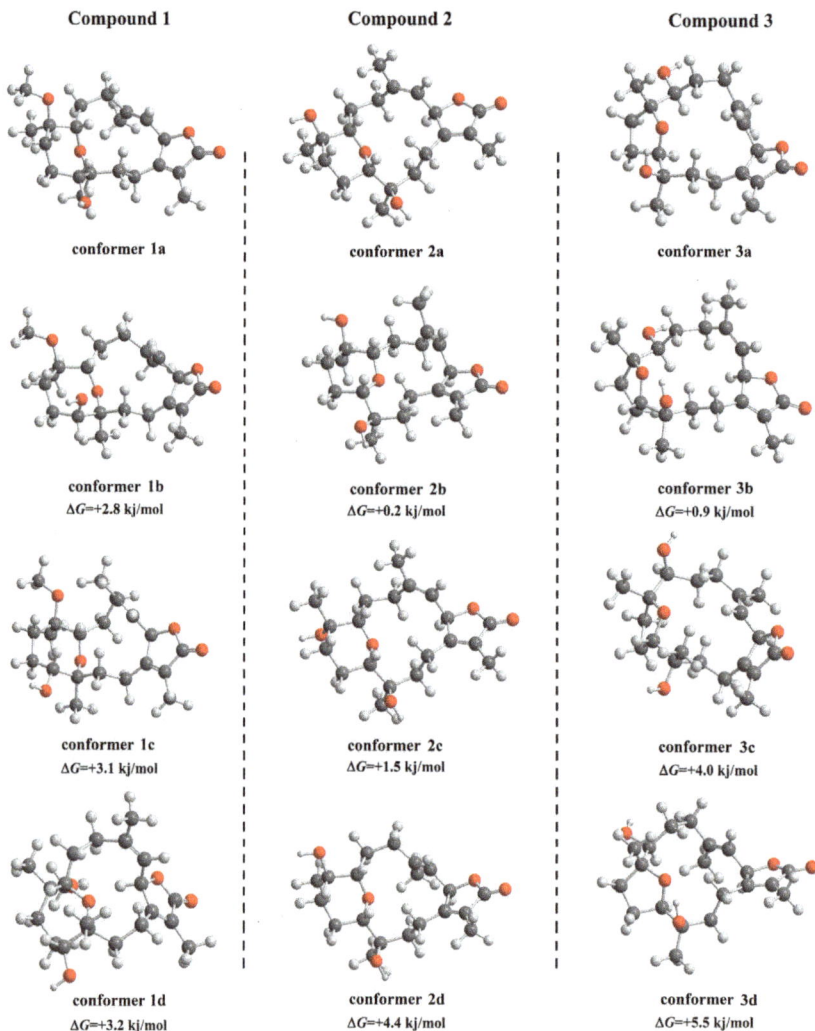

Compound 1	Compound 2	Compound 3
conformer 1a	conformer 2a	conformer 3a
conformer 1b ΔG=+2.8 kj/mol	conformer 2b ΔG=+0.2 kj/mol	conformer 3b ΔG=+0.9 kj/mol
conformer 1c ΔG=+3.1 kj/mol	conformer 2c ΔG=+1.5 kj/mol	conformer 3c ΔG=+4.0 kj/mol
conformer 1d ΔG=+3.2 kj/mol	conformer 2d ΔG=+4.4 kj/mol	conformer 3d ΔG=+5.5 kj/mol

Figure 6. Optimized confirmers for **1–3**, as well as the next three higher free-energy conformers.

2.2. Anti-Proliferative Activity against Cancer Lines

Isolated compounds **1–8** were evaluated for their anti-proliferative activity against three human tumor cell lines originating from the lung (A549), colon (Caco-2), and liver (HepG2) tissue based on an MTT reduction assay. The treatment of a human lung tumor cell line revealed differential anti-proliferative effects (Table 2). Among all tested compounds, sardisterol (**8**) had the most potent effect on A549 cells with a concentration-dependent loss of cell proliferation compared to a DMSO solvent control. Sardisterol was isolated for the first time from the marine soft coral *S. digitatum* [23]. To the best of our knowledge, this is the first report of anti-proliferative activity for **8** against tumor cells (Table 2 and Figure S25). The treatment of HepG2 cells with increasing concentrations of **1–8** revealed differential anti-proliferative potential (Table 2 and Figure S27). Compounds **3** and **8** showed a moderate inhibition (IC_{50} = 53.8 and 56.8 μM, respectively). Compounds **1, 3, 5, 6**, and **7** exhibited IC_{50} values between 63.1 and 98.6 μM.

Table 2. IC_{50} values * of tested compounds against A549, Caco-2, and HepG2 cells.

Compound	A549 IC_{50}(μM)	Caco-2 IC_{50}(μM)	HepG2 IC_{50}(μM)
1	50.1	>100	98.6
2	76.4	>100	>100
3	50.8	>100	53.8
4	91.5	>100	>100
5	62.2	>100	79.3
6	37.0	79.2	70.2
7	43.6	99.2	63.1
8	27.3	>100	56.8
Doxorubicin HCl	0.62	1.40	2.10

* IC_{50} values were obtained by fitting the concentration-response curve to non-linear regression. model on GraphPad
®Prism software v 6.0.

2.3. Molecular Docking Studies

The epidermal growth factor receptor (EGFR) is a tyrosine kinase receptor that is overexpressed in many tumor cell types and reported as a cause for some non-small-cell lung carcinomas [24]. Based on the hypothesis that anti-proliferation activity with the A549 cell line is associated with EGFR inhibition [25], receptor binding to an ATP binding site domain of EGFR kinase by isolated metabolites was examined. Doxorubicin, a known effector molecule that binds to the ATP binding domain of EGFR, was initially used as a positive control. The ability of **8** (the most active metabolite) and **4** (the least active) were examined using molecular docking simulations. Simulations were conducted with AutoDock 4.2 software with an initial assessment of docking done by comparing the docked pose of the co-crystallized inhibitor ligand erlotinib with the experimental pose (PDB code: 1M17). To further provide a predictive picture as to where the compounds are positioned in terms of direct interactions with EGFR, docking scores for the newly isolated natural products are provided and two other EGFR inhibitors, afatinib and gefitinib, as well as the positive control doxorubicin (Table S2). Per the predicted binding poses, AutoDock accurately reproduced the crystal structure observed binding modes of erlotinib, afatinib, and gefitinib (Figure 7a–c). To reveal the binding features and possible interactions, compounds were subjected to molecular docking simulation followed by AMBER-based molecular mechanical minimization and MM/GBSA binding energy calculations. The calculated binding energies are in good agreement with experimental data, with a correlation coefficient R^2 of 0.96 (Table S2, Figure S28). Per the minimized doxorubicin complex (Figure 7d), the highest potency of doxorubicin (ΔGMM/GBSA = −54.72 kcal/mol) may be attributed to its ability to form five hydrogen bonds with Lys_{721}, Thr_{766}, Met_{769}, Thr_{830}, and Asp_{831}, with bond lengths of 1.92, 1.76, 1.82, 1.92, and 1.94 Å, respectively. For **6–8**, ligand-receptor interactions are stabilized by two hydrogen bonds with the active site (Figure 7e–g), resulting in a stronger binding energy value of −41.18 kcal/mol in the case of **8**. For **4**, the lower binding energy may be due to the formation of only one hydrogen bond with the active site Lys_{721} (Figure 7h). Several van der Waals and hydrophobic interactions were observed between the added marine ligand and amino acids in the active site including Leu_{865}, Leu_{694}, and Val_{702}. To assess the stability of the predicted ligand-receptor complex, a short molecular dynamics simulation of 2.5 ns was performed for **8** in the complex with the EGFR receptor. The corresponding hydrogen bond distance between **8** and a carboxylate oxygen atom of Asp_{776} was then measured and averaged over the simulation time (Figure S29). According to the calculation, **8** can be considered stable, with an average hydrogen bond length of 2.16 Å with Asp_{776}.

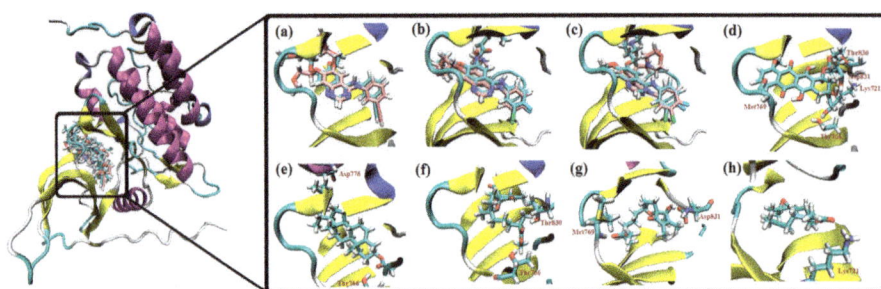

Figure 7. Crystal structure (in cyan) and predicted docking pose (in pink) of (**a**) erlotinib; (**b**) afatinib; and (**c**) gefitinib; and AMBER-based minimized docked structures of (**d**) doxorubicin; (**e**) **8**; (**f**) **7**; (**g**) **6**; and (**h**) **4** with EGFR kinase domain (PDB code: 1M17).

3. Experimental Section

3.1. General Experimental Procedures

Specific rotation was measured with a JASCO P-2200 polarimeter (JASCO Corporation, Tokyo, Japan) and the IR spectra were collected on a JASCO FT/IR-6300 spectrometer (JASCO Corporation, Tokyo, Japan). HR-ESI-FT-MS was carried out using a Thermo Fisher Scientific LTQ Orbitrap XL mass spectrometer (Waltham, MA, USA) at the Natural Science Center for Basic Research and Development (N-BARD), Hiroshima University. The ^{1}H (600 MHz) and ^{13}C (150 MHz) NMR spectra were recorded on a JEOL JNM-ECA 600 spectrometer (JEOL Ltd, Tokyo, Japan) with tetramethylsilane as an internal standard. Purification was run on a Shimadzu HPLC system equipped with a RID-10A refractive index detector and compound separation was performed on YMC-Pack ODS-A (YMC CO., LTD., Tokyo, Japan, 250 × 4.6 mm i.d., 5 μm) and (250 × 10 mm i.d., 5 μm) columns for analytical and preparative separation, respectively. Chromatography separation included normal-phase Silica gel 60 (230–400 mesh, Merck, Darmstadt, Germany), which was used for column chromatography. Pre-coated silica gel plates (Kieselgel 60 F_{254}, 0.25 mm, Merck, Darmstadt, Germany) were used for TLC analyses. Spots were visualized by heating after spraying with 10% H_2SO_4.

3.2. Animal Material

Soft coral *Sarcophyton ehrenbergi* was collected from the Egyptian Red Sea off the coast of Hurghada in March 2015. The soft coral was identified by M Al-Hammady with a voucher specimen (03RS27) deposited in the National Institute of Oceanography and Fisheries, marine biological station, Hurghada, Egypt.

3.3. Extraction and Separation

Frozen soft coral (5.2 kg, total wet weight) was chopped into small pieces and extracted with methylene chloride/methanol (1:1) at room temperature (5 L × 5 times). The combined extracts were concentrated in vacuo to a brown gum. The dried material (218 g) was subjected to gravity chromatography in a silica gel column (6 × 120 cm) eluting with *n*-hexane (3000 mL), followed by a gradient of *n*-hexane-CH_2Cl_2 up to 100% CH_2Cl_2 and CH_2Cl_2–MeOH up to 50% MeOH (3000 mL each of the solvent mixture). The *n*-hexane/CH_2Cl_2 (1:1) fraction (2.2 g) eluted with *n*-hexane/EtOAc (6:1) was subjected to silica gel column separation. Fractions were obtained and combined into two main sub-fractions, A and B, according to a TLC profile. Sub-fraction A was re-purified by reversed-phase HPLC using MeOH/H_2O (6.5:3.5), 3.5 mL/min, to afford **1** (6.1 mg, t_R = 27 min), **4** (11.6 mg, t_R = 23 min), and **5** (11.6 mg, t_R = 21 min). Sub-fraction B was re-purified by reversed-phase HPLC using MeOH/H_2O (3:2), 3.5 mL/min, to afford **2** (10 mg, t_R = 28 min) and **3** (7.5 mg, t_R = 29.5 min).

The *n*-hexane/CH_2Cl_2 (1:2) fraction (1.4 g) was subjected to silica gel column chromatography eluted by *n*-hexane/EtOAc (5:1) that afforded the main sub-fraction C. Sub-fraction C was re-purified by reversed-phase HPLC using MeOH/H_2O (1:1), 3 mL/min, to afford **6** (9.0 mg, t_R = 31 min), **7** (11.2 mg, t_R = 32 min), and **8** (14.1 mg, t_R = 37 min).

2S,16:7R,12S-Diepoxy-11α-hydroxy-8β-methoxy-16-keto-cembra-1E,3E-diene (*Sarcoehrenbergilid A*, **1**): white crystals; $[\alpha]_{25}^D$ −6.9 (*c* 0.0023, $CHCl_3$); FT-IR (KBr) ν_{max}: 3450, 2933, 1745, 1455, and 1220 cm^{-1}; ^{1}H and ^{13}C NMR data, see Table 1; HRESI-FTMS *m/z* 387.2142 [100, (M + Na)$^+$]; (calcd. 364.2250, for $C_{21}H_{32}O_5$).

2S,16:8S,11S-Diepoxy-7β,12α-dihydroxy-16-keto-cembra-1E,3E-diene (*Sarcoehrenbergilid B*, **2**): white powder; $[\alpha]_{25}^D$ −6.6 (*c* 0.003, $CHCl_3$); FT-IR (KBr) ν_{max}: 3447, 2933, 1747, 1457, and 1219 cm^{-1}; ^{1}H and ^{13}C NMR data, see Table 1; HRESI-FTMS *m/z* 373.1986 [100, (M + Na)$^+$]; (calcd. 350.2093, for $C_{20}H_{30}O_5$).

2S,16:7R,11R-Diepoxy-8β,12β-dihydroxy-16-keto-cembra-1E,3E-diene (*Sarcoehrenbergilid C*, **3**): white amorphous powder; $[\alpha]_{25}^D$ −3.7 (*c* 0.0027, $CHCl_3$); FT-IR (KBr) ν_{max}: 3447, 2933, 1747, 1451, and 1231 cm^{-1}; ^{1}H and ^{13}C NMR data, see Table 1; HRESI-FTMS *m/z* 373.1985 [100, (M + Na)$^+$]; (calcd. 350.2093 for $C_{20}H_{30}O_5$).

X-ray Crystallography Data

Data collection was performed with a Bruker SMART-APEX II ULTRA CCD area detector with graphite monochromated Cu Kα radiation (λ = 1.54178 Å) at the Center for Analytical Instrumentation, Chiba University, Japan. The structure was solved by direct methods using SHELXS-97 [26]. Refinements were performed with SHELXL-2013 [27] using full-matrix least squares on F^2. All non-hydrogen atoms were refined anisotropically. All hydrogen atoms were placed in idealized positions and refined as riding atoms isotropically. Crystal data: $C_{21}H_{32}O_5$, *M* = 364.46, monoclinic, crystal size, 0.30 × 0.30 × 0.10 mm^3, Space group *P*21, *Z* = 2, crystal cell parameters *a* = 5.74330 (10) Å, *b* = 16.8653 (3) Å, *c* = 10.2383 (2) Å, α = 90°, β = 95.7012 (7)°, γ = 90°, *V* = 986.80 (3) Å^3, *F*(000) = 396, *Dc* = 1.227 Mg/m^3, *T* = 173 K, 12540 reflections measured, 3480 independent reflections [$R_{(int)}$ = 0.0216], final *R* indices [*I* > 2.0σ(*I*)], R_1 = 0.0307, wR_2 = 0.0844; final *R* indices (all data), R_1 value = 0.0308, wR_2 = 0.0845, Flack parameter [24]: 0.09 (3). CCDC-1532555 contains the supplementary crystallographic data for this paper. The data can be obtained free of charge from The Cambridge Crystallographic Data Centre via http://www.ccdc.cam.ac.uk/conts/retrieving.html (or from the CCDC, 12 Union Road, Cambridge CB2 1EZ, UK; Fax: +44 1223 336033; E-mail: deposit@ccdc.cam.ac.uk).

3.4. Cell Culture

All materials and reagents for the cell cultures were purchased from Lonza (Verviers, Belgium). Human cancer cell lines of non-small cell lung adenocarcinoma (A549), colon adenocarcinoma (Caco-2), and hepatocellular carcinoma (HepG2) (ATCC®) were maintained as monolayer culture in Dulbecco's modified Eagle's medium (DMEM) supplemented with 10% FBS, 4 mM L-glutamine, 100 U/mL penicillin, and 100 μg/mL streptomycin sulfate. Monolayers were passaged at 70–90% confluence using a trypsin-EDTA solution. All cell incubations were maintained in a humidified CO_2 incubator with 5% CO_2 at 37 °C.

3.5. Cell Proliferation Assay

Anti-proliferative studies were performed using a modified MTT (3-[4,5]-2,5-diphenyltetrazolium bromide) assay based on a previously published method [28,29]. Appropriate cell densities of exponentially growing A549, Caco-2, or HepG2 cells (5000–10000 cells/well) were seeded onto 96-well plates. After a 24 h incubation period with 5% CO_2 at 37 °C, stock test compounds (**1–8**) dissolved in dimethyl sulfoxide (DMSO) were added at concentrations of 100, 50, 25, 12.5, and 6.25 μM in culture medium (final DMSO concentration in medium = 0.1%, by volume). After 48 h of incubation, MTT solution in PBS (5 mg/mL) was added to each well, after which the incubation was resumed for a further 90 min. The formation of intracellular formazan crystals (mitochondrial reduction product of MTT) was confirmed by a phase contrast microscopic examination. At the end of the incubation period, the medium was removed, and 100 μL of DMSO was added to each well to dissolve formed formazan crystals with shacking for 10 min (200 rpm). Dissolved crystals were quantified by reading the absorbance at 492 nm (OD) on a microplate reader (Sunrise™ microplate reader, Tecan Austria Gmbh, Grödig, Austria) and were used as a measure of cell proliferation.

3.6. Anti-Proliferation Quantitative Analysis

Cell proliferation was determined by comparing the average OD values of the control wells with those of the samples (quadrate to octuplet treatments), both represented as % proliferation [control proliferation (0.1% DMSO only) = 100%]. The IC_{50} values (concentration of sample causing 50% loss of cell proliferation of the vehicle control) were calculated using the concentration-response curve fit to the non-linear regression model using GraphPad Prism® v6.0 software (GraphPad Software Inc., San Diego, CA, USA).

3.7. Computational Methodology

3.7.1. Density Functional Theory Calculations

The conformational structures for **1–3** were generated using Omega2 software (version 2.5.1.4, OpenEye Scientific Software, Santa Fe, NM, USA) [30]. In the conformational search, the energy window value was set to 10 kcal/mol and all stereogenic centers were considered; other parameters were set to default. The geometry of each generated conform was then energetically optimized at the B3LYP/6-31G* level of theory using Gaussian09 software (Revision E.01, Gaussian, Inc., Wallingford, CT, USA) [31]. All optimized conformers were subjected to a vibrational frequency calculation to confirm the minimum energy states and the corresponding free energies were obtained. The relative Boltzmann population of each conformer was then valued at 298 K. Optical rotations were predicted at the same level of theory.

3.7.2. Molecular Docking Studies

Molecular docking in the ATP binding site of EGFR kinase domain was performed using AutoDock 4.2 software (version 4.2, The Scripps Research Institute, La Jolla, CA, USA) [32]. The crystal structure of the EGFR kinase domain complexed with erlotinib (PDB code: 1M17 [33]) was taken as the template for all docking calculations. Water molecules were deleted and all missing hydrogen atoms were added based on the protonation state of the protein. The receptor pdbqt file was then prepared according to AutoDock protocol [34]. The grid center was centered on an erlotinib inhibitor, and the grid box size was set to $60 \times 60 \times 60$ points with a grid spacing of 0.375 Å.

The number of Autodock GA runs was set to 50 and maximum number of energy evaluations was set to 2,500,000. Other AutoDock parameters were set to their default values. 3D structures were constructed and minimized using an MMFF94S force field with the help of SZYBKI software (version 1.9.0.3, OpenEye Scientific Software, Santa Fe, NM, USA) [35]; atomic charges were assigned using a gasteiger method. Prior to the binding energy calculation, all docked complexes were minimized using AMBER14 software (version 14, University of California, San Francisco, CA, USA) [36]. The studied compounds and receptor were described by the general AMBER force field (GAFF) [37] and AMBER force field 14SB [38], respectively. The atomic partial charges of the studied inhibitors were evaluated using the restrained electrostatic potential (RESP) approach at the HF/6-31G* level. For minimization, the truncated Newton linear conjugate gradient method with LBFGS preconditioning was used. The convergence criterion for the energy gradient was 10^{-9} kcal/mol·Å. A cutoff value of 999 Å and a generalized Born solvent model were used. On the basis of the minimized complex structures, the binding energies were calculated using the molecular mechanical–generalized Born surface area (MM-GBSA) approach.

The estimated MM-GBSA binding energies were correlated to experimental binding energies, which were calculated based on A549 inhibitory constants using the following equation: $\Delta G_{exp} = RT \ln(IC_{50})$, where IC_{50} in μM and $T = 298.15$ K. To assess the compound **8** stability inside the EGFR active site, a short molecular mechanical simulation of an 8-EGFR complex was first performed using the AMBER software. For the MD simulation, the complex was neutralized and solvated with TIP3P water molecules. The complex was then energetically minimized, heated gradually over a period of 50 ps to 300 K, and equilibrated for 500 ps. The data were then collected over a 2.5 ns simulation with a time step of 2 fs and the SHAKE option to constrain all bonds involving hydrogen atoms was used.

4. Conclusions

Three new (**1**–**3**) and five previously reported (**4**–**8**) terpenoids were isolated and chemically characterized from the Red Sea soft coral *S. ehrenbergi*. The eight identified compounds exhibited differential antiproliferative potential against three human cancer cell lines, with lung A549 cell being the most sensitive to compound treatment. The present study establishes *S. ehrenbergi*as as a new source of sardistrol and a possible antiproliferative candidate against lung cancer. Molecular docking studies are consistent with the binding of **8** to the EGFR kinase domain and the inhibition of cell growth. Molecular docking studies supported high inhibitory activity for **8** versus **6** or **7** with the EGFR kinase domain.

Supplementary Materials: The following are available online at www.mdpi.com/1660-3397/15/6/192/s1, Figures S1–S29: HR-ESI-MS, 1D, and 2D NMR spectra of compounds **1**–**3**, Table S1: Conformer relative free energies with respect to the most stable form, Table S2: Calculated auto-dock and MM/GBSA binding energies (ΔG) for the most and least active compounds complexed with the EGFR kinase domain.

Acknowledgments: This work was financially supported by National Research Centre, Egypt, and the Welch Foundation (D-1478).

Author Contributions: M.-E.F. Hegazy, A.I. Elshamy and T.A. Mohamed contributed to the extraction, isolation, purification, identification, and manuscript preparation. A.R. Hamed contributed to the cytotoxicity experiments, analysis, and manuscript preparation. M.A.A. Ibrahim contributed to the computational studies, analysis, and manuscript preparations. S. Ohta and P.W. Paré contributed to the structure elucidation, guiding experiments, analyses, and manuscript preparations. M.-E.F. Hegazy was the project leader, organizing and guiding the experiments, structure elucidation, and manuscript writing.

Conflicts of Interest: The authors declare no conflict of interest.

References

1. Yang, B.; Zhou, X.-F.; Lin, X.-P.; Liu, J.; Peng, Y.; Yang, X.-W.; Liu, Y. Cembrane diterpenes chemistry and biological properties. *Curr. Org. Chem.* **2012**, *16*, 1512–1539. [CrossRef]
2. Kobayashi, M.; Nakagawa, T.; Mitsuhashi, H. Marine terpenes and terpenoids. I. Structures of four cembrane-type diterpenes: Sarcophytol-A, sarcophytol-A acetate sarcophytol-B, and sarcophytonin-A, from the soft coral, *Sarcophyton glaucum*. *Chem. Pharm. Bull.* **1979**, *27*, 2382–2387. [CrossRef]
3. Yang, B.; Liu, J.; Wang, J.; Liao, S.; Liu, Y. Handbook of anticancer drugs from marine origin. In *Cytotoxic Cembrane Diterpenoids*; Kim, S.-K., Ed.; Springer International Publishing: Cham, Switzerland, 2015.
4. Abou El-Ezz, R.; Ahmed, S.; Radwan, M.; Ayoub, N.; Afifi, M.; Ross, S.; Szymanski, P.; Fahmy, H.; Khalifa, S. Bioactive cembranoids from the Red Sea soft coral *Sarcophyton glaucum*. *Tetrahedron Lett.* **2013**, *54*, 989–992. [CrossRef]
5. Liang, L.; Guo, Y. Terpenes from the soft corals of the genus *Sarcophyton*: Chemistry and biological activities. *Chem. Biodivers.* **2013**, *10*, 2161–2196. [CrossRef] [PubMed]
6. Hegazy, M.E.; El-Beih, A.A.; Moustafa, A.Y.; Hamdy, A.A.; Alhammady, M.A.; Selim, R.M.; Abdel-Rehim, M.; Paré, P.W. Cytotoxic cembranoids from the Red Sea soft coral *Sarcophyton glaucum*. *Nat. Prod. Commun.* **2011**, *6*, 1809–1812. [PubMed]

7.   Hegazy, M.E.; Mohamed, T.A.; Abdel-Latif, F.A.; Alsaid, M.; Shahat, A.A.; Paré, P.W. Trochelioid A and B, new cembranoid diterpenes from the Red Sea soft coral *Sarcophyton trocheliophorum*. *Phytochem. Lett.* **2013**, *6*, 383–386. [CrossRef]

8.   Elkhateeb, A.; El-Beih, A.; Gamal-Eldeen, A.; Alhammady, M.; Ohta, S.; Paré, P.; Hegazy, M.E. New terpenes from the Egyptian soft coral *Sarcophyton ehrenbergi*. *Mar. Drugs* **2014**, *12*, 1977–1986. [CrossRef] [PubMed]

9.   Yan, X.-H.; Li, Z.-Y.; Guo, Y.-W. Further new cembranoid diterpenes from the Hainan soft coral *Sarcophyton latum*. *Helv. Chim. Acta* **2007**, *90*, 1574. [CrossRef]

10.   Yan, X.-H.; Feng, L.-Y.; Guo, Y.-W. Further new cembrane diterpenes from the Hainan soft coral *Sarcophyton latum*. *Chin. J. Chem.* **2008**, *26*, 150–152. [CrossRef]

11.   Rodriguez, A.D.; Shi, J.G. The first cembrane-pseudopterane cycloisomerization. *J. Org. Chem.* **1998**, *63*, 420–421. [CrossRef] [PubMed]

12.   Qin, S.; Huang, H.; Guo, Y.W. A new cembranoid from the Hainan soft coral *Sinularia* sp. *J. Asian Nat. Prod. Res.* **2008**, *10*, 1075–1079. [CrossRef] [PubMed]

13.   Hegazy, M.E.F.; Mohamed, T.A.; Alhammady, M.A.; Shaheen, A.M.; Reda, E.H.; Elshamy, A.I.; Aziz, M.; Paré, P.W. Molecular architecture and biomedical leads of terpenes from Red Sea marine invertebrate. *Mar. Drugs* **2015**, *13*, 3154–3181. [CrossRef] [PubMed]

14.   Torre, L.A.; Bray, F.; Siegel, R.L.; Ferlay, J.; Lortet-Tieulent, J.; Jemal, A. Global cancer statistics, 2012. *CA Cancer J. Clin.* **2015**, *65*, 87–108. [CrossRef] [PubMed]

15.   Society, A.C. Cancer Facts & Figures. 2016. Available online: http://www.cancer.org/acs/groups/content/ @research/documents/document/acspc-047079.pdf (accessed on 15 June 2017).

16.   Hegazy, M.E.F.; Mohamed, T.A.; Elshamy, A.I.; Alhammady, M.A.; Ohta, S.; Paré, P. Casbane Diterpenes from Red Sea Coral *Sinularia polydactyla*. *Molecules* **2016**, *21*, 308. [CrossRef] [PubMed]

17.   Parsons, S.; Flack, H.D.; Wagner, T. Use of intensity quotients and differences in absolute structure refinement. *Acta Cryst.* **2013**, *69*, 249–259. [CrossRef] [PubMed]

18.   Parsons, S.; Flack, H. Precise absolute-structure determination in light-atom crystals. *Acta Cryst.* **2004**, *60*, s61. [CrossRef]

19.   Sawant, S.; Sylvester, P.; Avery, M.; Desai, P.; Youssef, D.; El Sayed, K. Bioactive rearranged and halogenated semisynthetic derivatives of the marine natural product sarcophine. *J. Nat. Prod.* **2004**, *67*, 2017–2023. [CrossRef] [PubMed]

20.   Xi, Z.; Bie, W.; Chen, W.; Liu, D.; Ofwegen, L.V.; Proksch, P.; Lin, W. Sarcophyolides B–E, new cembranoids from the soft coral *Sarcophyton elegans*. *Mar. Drugs* **2013**, *11*, 3186–3196. [CrossRef] [PubMed]

21.   Su, J.-H.; Lu, Y.; Lin, W.-Y.; Wang, W.-H.; Sung, P.-J.; Sheu, J.-H. A cembranoid, trocheliophorol, from the cultured soft coral *Sarcophyton trocheliophorum*. *Chem. Lett.* **2010**, *39*, 172–173. [CrossRef]

22.   Yang, B.; Wei, X.; Huang, J.; Lin, X.; Liu, J.; Liao, S.; Wang, J.; Zhou, X.; Wang, L.; Liu, Y. Sinulolides A–H, new cyclopentenone and butenolide derivatives from soft coral *Sinularia* sp. *Mar. Drugs* **2014**, *12*, 5316–5327. [CrossRef] [PubMed]

23.   Su, J.; Yang, R.; Zeng, L. Sardisterol, a new polyhydroxylated sterol from the soft coral *Sarcophyton digitatum* Moser. *Chin. J. Chem.* **2001**, *19*, 515–517. [CrossRef]

24.   Dutta, P.R.; Maity, A. Cellular responses to EGFR inhibitors and their relevance to cancer therapy. *Cancer Lett.* **2007**, *254*, 165–177. [CrossRef] [PubMed]

25.   Sun, J.; Wang, X.-Y.; Lv, P.-C.; Zhu, H.-L. Discovery of a series of novel phenylpiperazine derivatives as EGFR TK inhibitors. *Sci. Rep.* **2015**, *5*, 13934. [CrossRef] [PubMed]

26.   Sheldrick, G.M. *SHELXS-97: Program for Crystal Structure Solution*; University of Göttingen: Göttingen, Germany, 1997.

27.   Sheldrick, G.M. *SHELXL-2013: Program for Crystal Structure Refinement*; University of Göttingen: Göttingen, Germany, 2013.

28.   Mosmann, T. Rapid colorimetric assay for cellular growth and survival: Application to proliferation and cytotoxicity assays. *J. Immunol. Methods* **1983**, *65*, 55–63. [CrossRef]

29.   Dutta, A.; Bandyopadhyay, S.; Mandal, C.; Chatterjee, M. Development of a modified MTT assay for screening antimonial resistant field isolates of Indian visceral leishmaniasis. *Parasitol. Int.* **2005**, *54*, 119–122. [CrossRef] [PubMed]

30. Hawkins, P.C.D.; Skillman, A.G.; Warren, G.L.; Ellingson, B.A.; Stahl, M.T. Conformer Generation with OMEGA: Algorithm and Validation Using High Quality Structures from the Protein Databank and Cambridge Structural Database. *J. Chem. Inf. Model.* **2010**, *50*, 572. [CrossRef] [PubMed]

31. Frisch, M.J.; Trucks, G.W.; Schlegel, H.B.; Scuseria, G.E.; Robb, M.A.; Cheeseman, J.R.; Scalmani, G.; Barone, V.; Mennucci, B.; Petersson, G.A.; et al. *Gaussian09*; Revision E.01; Gaussian Inc.: Wallingford, CT, USA, 2009.

32. Morris, G.M.; Huey, R.; Lindstrom, W.; Sanner, M.F.; Belew, R.K.; Goodsell, D.S.; Olson, A.J. AutoDock4 and AutoDockTools4: Automated docking with selective receptor flexibility. *J. Comput. Chem.* **2009**, *30*, 2785–2791. [CrossRef] [PubMed]

33. Stamos, J.; Sliwkowski, M.X.; Eigenbrot, C. Structure of the Epidermal Growth Factor Receptor Kinase Domain Alone and in Complex with a 4-Anilinoquinazoline Inhibitor. *J. Biol. Chem.* **2002**, *277*, 46265–46272. [CrossRef] [PubMed]

34. Forli, S.; Huey, R.; Pique, M.E.; Sanner, M.F.; Goodsell, D.S.; Olson, A.J. Computational protein-ligand docking and virtual drug screening with the AutoDock suite. *Nat. Protoc.* **2016**, *11*, 905–919. [CrossRef] [PubMed]

35. *SZYBKI*, version 1.9.0.3; OpenEye Scientific Software: Santa Fe, NM, USA, 2016.

36. Case, D.; Babin, V.; Berryman, J.; Betz, R.; Cai, Q.; Cerutti, D.; Cheatham, T.; Darden, T.; Duke, R.; Gohlke, H.; et al. *Amber 14*; University of California: San Francisco, CA, USA, 2014.

37. Wang, J.; Wolf, R.M.; Caldwell, J.W.; Kollman, P.A.; Case, D.A. Development and testing of a general amber force field. *J. Comput. Chem.* **2004**, *25*, 1157–1174. [CrossRef] [PubMed]

38. Maier, J.A.; Martinez, C.; Kasavajhala, K.; Wickstrom, L.; Hauser, K.E.; Simmerling, C. ff14SB: Improving the Accuracy of Protein Side Chain. *J. Chem. Theory Comput.* **2015**, *11*, 3696–3713. [CrossRef] [PubMed]

marine drugs

MDPI

Article

Cytotoxicity of Endoperoxides from the Caribbean Sponge *Plakortis halichondrioides* towards Sensitive and Multidrug-Resistant Leukemia Cells: Acids vs. Esters Activity Evaluation

Tanja Schirmeister [1,*], Swarna Oli [1], Hongmei Wu [1], Gerardo Della Sala [2], Valeria Costantino [2], Ean-Jeong Seo [1] and Thomas Efferth [1]

[1]   Institute of Pharmacy and Biochemistry, Johannes Gutenberg University Mainz, Staudinger Weg 5, 55128 Mainz, Germany; swarnoli@uni-mainz.de (S.O.); hongmei@uni-mainz.de (H.W.); seo@uni-mainz.de (E.-J.S.); efferth@uni-mainz.de (T.E.)
[2]   The NeaNat Group, Dipartimento di Farmacia, Università degli Studi di Napoli Federico II, via D. Montesano 49, 80131 Napoli, Italy; gerardo.dellasala@unina.it (G.D.S.); valeria.costantino@unina.it (V.C.)
*   Correspondence: schirmei@uni-mainz.de; Tel.: +49-6131-392-5742

Academic Editor: Friedemann Honecker
Received: 15 December 2016; Accepted: 16 February 2017; Published: 3 March 2017

Abstract: The 6-epimer of the plakortide H acid (**1**), along with the endoperoxides plakortide E (**2**), plakortin (**3**), and dihydroplakortin (**4**) have been isolated from a sample of the Caribbean sponge *Plakortis halichondrioides*. To perform a comparative study on the cytotoxicity towards the drug-sensitive leukemia CCRF-CEM cell line and its multi-drug resistant subline CEM/ADR5000, the acid of plakortin, namely plakortic acid (**5**), as well as the esters plakortide E methyl ester (**6**) and 6-epi-plakortide H (**7**) were synthesized by hydrolysis and Steglich esterification, respectively. The data obtained showed that the acids (**1**, **2**, **5**) exhibited potent cytotoxicity towards both cell lines, whereas the esters showed no activity (**6**, **7**) or weaker activity (**3**, **4**) compared to their corresponding acids. Plakortic acid (**5**) was the most promising derivative with half maximal inhibitory concentration (IC_{50}) values of ca. 0.20 µM for both cell lines.

Keywords: Caribbean sponge; plakortide; endoperoxide; leukemia; multi-drug resistant leukemia; cytotoxicity

1. Introduction

Marine organisms are excellent sources of novel skeletons ranging from small terpene molecules [1,2], mixed polyketide-peptide biogenesis [3,4], to more complex carbohydrate-based scaffolds [5,6]. Many of these novel skeletons [7] have been tested for their possible role as lead compounds in the search for new drugs for various diseases. Among the different classes, endoperoxides such as the famous artemisinin from *Artemisia annua* L. are well-known for their bioactivity. The Chinese scientist Youyou Tu isolated artemisinin and described its antimalarial activity in the 1970s. She was honoured with the Nobel Prize for Physiology or Medicine in 2015 [8]. Artemisinin and its derivatives are also active against various cancer cell lines, especially against leukemia and colon cancer [9,10]. The first long-term treatment of cancer patients with artesunate in combination with standard chemotherapy has been described [11]. In 2009, the combined effects of artesunate and rituximab on malignant B-cells were reported [12]. Clinical pilot phase I/II trials in veterinary tumors and human cancer patients demonstrated that the artemisinin derivative artesunate possesses clinical anticancer activity at tolerable side effects [13–15]. It can be speculated that not only artemisinin-type drugs, but also other endoperoxides may reveal anticancer activity. This hypothesis is substantiated by reports on the cytotoxicity of natural and synthetic endoperoxides towards tumor cell lines [16–25]. Endoperoxides are, therefore, worth investigating to unravel their full potential as anticancer drug leads. The Caribbean sponge *Plakortis halichondrioides* produces endoperoxides which were assumed to be synthesized by the polyketide pathway [26,27]. Similar to artesunate, these metabolites did not only display antimalarial activity, but also cytotoxic activity against several tumor cell lines [28–30]. From a sample of this sponge, we isolated plakortide E (**2**, Figure 1) and found that it was also active against trypanosomes [31]. Here, we report the cytotoxicity towards the drug-sensitive leukemia CCRF-CEM cell line (human Caucasian acute lymphoblastic leukemia, childhood T acute lymphoblastic leukemia) and its multi-drug resistant subline CEM/ADR5000 (multi-drug resistant CCRF cell line) (Table 2), of seven derivatives (Figure 1): the 6-epimer of the plakortide H acid (**1**) [32,33] along with the endoperoxides plakortide E (**2**), plakortin (**3**) [34–36], and dihydroplakortin (**4**) [36,37] that have been isolated from a sample of the Caribbean sponge *Plakortis halichondrioides*. In addition, the acid of plakortin, namely plakortic acid (**5**) [38,39], as well as the esters plakortide E methyl ester (**6**) [40,41] and the ester 6-epi-plakortide H (**7**) were synthesized by hydrolysis (plakortic acid) and Steglich esterification (plakortide E methyl ester and 6-epi-plakortide H), respectively, to perform a comparative study. There are some discrepancies within the literature concerning the nomenclature of plakortides and their esters: According to reference [22] plakortide I is the acid of the methyl ester plakortide H. Also reference [27] and the reference [32] term the methyl ester plakortide H. In contrast, the reference [38] describes plakortide H as the respective acid and plakortide I as its 11,12-dihydro derivative. In the present manuscript, we refer to plakortide H as the methyl ester, and accordingly compound **1** is the 6-epimer of plakortide H acid, and compound **7** the 6-epimer of plakortide H. There are also discrepancies concerning the structure of plakortic acid: According to reference [20] the natural compound named plakortic acid is rather an epoxide than an endoperoxide. Reference [38] in contrast assigns the structure of the acid of plakortin to plakortic acid. In the present manuscript, we refer to plakortic acid **5** as the acid of plakortin **3**.

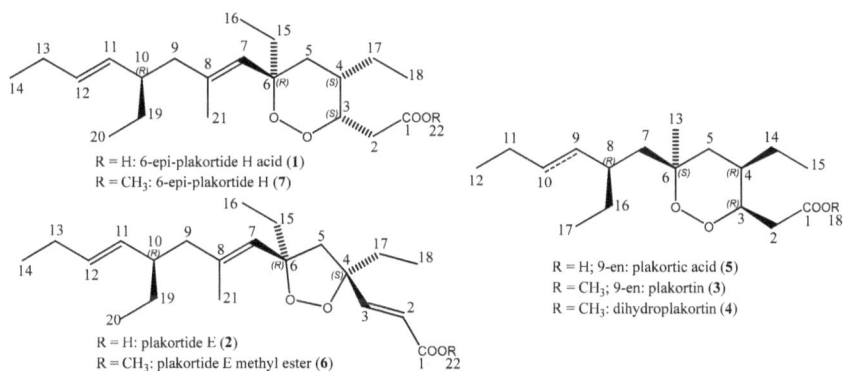

Figure 1. Structures of natural (**1**, **2**, **3**, **4**) and semi-synthetic (**5**, **6**, **7**) endoperoxides from a sample of the sponge *Plakortis halichondrioides*: 6-epi-plakortide H acid (**1**), plakortide E (**2**), plakortin (**3**), dihydroplakortin (**4**), plakortic acid (**5**), plakortide E methyl ester (**6**), and 6-epi-plakortide H (**7**).

2. Results

2.1. Isolation, Semi-Syntheses, and Identification of 6-Epi-Plakortide H Acid (**1**) and Its Methyl Ester 6-Epi-Plakortide H (**7**)

A sample of the sponge *Plakortis halichondrioides*, order Homosclerophorida, family Plakinidae, (640 g freeze-dried) was collected via scuba diving along the coast of Inagua Island (GPS coordinates

21°10.7684′ N 73°9.1608′ W) on 7 July 2013 at a depth of 30 m. After collection, the sample was unambiguously identified on board using a web-based photographic and taxonomic key [42]. The sample was immediately frozen and stored. A voucher sample with the reference no. 13/7/13 has been deposited at the Dipartimento di Farmacia, Università degli Studi Napoli "Federico II". For this study, the sponge tissue was cut into small pieces, lyophilized, and then sequentially extracted with cyclohexane, methylene chloride, and methanol solvents. The crude methylene chloride extract was subjected to column chromatography using a gradient solvent system starting with cyclohexane and changing gradually to methylene chloride, chloroform, and finally to methanol. Based upon thin layer chromatography (TLC) analysis the fractions were combined to yield six fractions I–VI (I-3.2 g, II-5.1 g, III-2.8 g, IV-4.3 g, V-6.9 g, VI-7.5 g). The fraction IV was subjected to preparative reversed-phase high performance liquid chromatography (RP-HPLC) chromatography to yield a fraction (termed **1mix**, 0.5005 g), which was identified as a mixture of several acidic compounds. The fraction was converted into an ester mixture (termed **7mix**) using the Steglich esterification procedure with methanol, dicyclohexylcarbodiimide (DCC), and 4-dimethylamino pyridine (DMAP). Then, the mixture of esters was purified using preparative RP-HPLC to yield a pure methyl ester (**7**), which eluted at 14 min as a pale yellow viscous oil. The ester which was later on identified as the methyl ester derivative of the 6-epi-plakortide H acid was hydrolyzed in THF/water (4:1; 10 mL) with LiOH (3 eq.). The residue obtained after acidic workup was further purified via preparative RP-HPLC to yield the pure acid (**1**). The structure of the compound was analyzed by ^{1}H, ^{13}C, correlation spectroscopy (COSY), and nuclear Overhauser exchange spectroscopy (NOESY) nuclear magnetic resonance (NMR) spectroscopy and mass spectrometry and, according to the literature [29,33] and NOESY data, the pure acid was identified as a diastereomer of plakortide H acid, namely the 6-epimer. In fact, the NOESY data and coupling constants are in agreement with those found for plakortides M and N [29] and are in agreement with the literature [33] and thus, the same configuration was also assumed for the isolated compound, namely the (R)-configuration at the 6-position and the (R)-configuration at the carbon atom 10 [33]. For coupling of H-3 (equatorial, eq.) and H-4(axial, ax.) a constant of *J* = 5.2 Hz was found. NOESY correlations (Figure 2) were observed between H-2 and H-5b(ax.), H-3(eq.) and H-4(ax.), H-4(ax.) and H-5a(eq.), H-4(ax.) and H-7, H5a(eq.) and H-7, and H-5b(ax.) and H-15 (Figure 2). This is only possible with an equatorial position of the ethyl moiety (i.e., (R)-configuration) at C-6. Thus, the absolute configuration was assigned as (6R,10R). NMR data for compound (**1**) are reported in Table 1, and the NMR data of methyl ester (**7**) are reported in the Supplementary Materials.

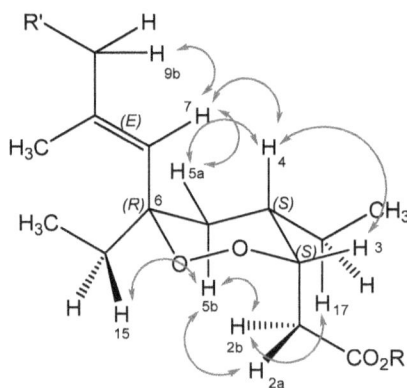

Figure 2. Selected nuclear Overhauser effect (NOE) correlations observed for 6-epi-plakortide H acid (**1**, R = H) and the methyl ester 6-epi-plakortide H (**7**, R = CH$_3$).

Table 1. ¹H Nuclear magnetic resonance NMR (600 MHz), ¹³C NMR (150 MHz), and nuclear Overhauser exchange spectroscopy (NOESY) spectral data for 6-epi-plakortide H acid (**1**) in CDCl₃.

Position	δ_C	Mult	δ_H	Mult	J in Hz	NOESY
1	177.06	C				
2	31.31	CH₂	3.07 (2a)	dd	15.9, 9.6	2b, 5b
			2.41 (2b)	dd	15.9, 3.4	2a, 2b, 17
3	78.64	CH	4.44	ddd	3.3, 5.2, 9.5	2a, 2b, 4
4	35.37	CH	2.09 [a]			3, 5a, 7
5	35.52	CH₂	1.61 [a] (5a)	m		4, 7, 5b
			1.26 [a] (5b)			2a, 2b, 5a, 15
6	84.49	C				
7	127.13	CH	5.12	s		4, 5a, 9b
8	137.58	C				
9	47.60	CH₂	2.06 [a]–1.94 [a]			7
10	42.60	CH	2.02 [a]			
11	133.14	CH	5.09	dd	15.1	
12	131.89	CH	5.35	dt	15.1, 6.2, 6.2	
13	25.77	CH₂	1.97 [a]			
14	14.15	CH₃	0.98	t	7.4	
15	32.58	CH₂	1.55	m		5b
16	7.78	CH₃	0.86	t	7.4	
17	25.12	CH₂	1.16 [a]			2b
18	11.12	CH₃	0.92	t	7.6	
19	28.05	CH₂	1.39	m		
			1.17 [a]			
20	11.78	CH₃	0.84	t	7.4	
21	17.04	CH₃	1.70	s		

Chemical shift values are in ppm relative to the residual peaks of CDCl₃ at 7.26 ppm (¹H), and 77.16 ppm (¹³C). Spectra were recorded at 25 °C. [a] Overlap with other signals. For the methyl ester **7**, the same NOE correlations were found.

2.2. Isolation and Identification of Plakortide E (**2**), Plakortin (**3**), and Dihydroplakortin (**4**)

The crude cyclohexane extract was subjected to chromatography on silica gel using the isocratic solvent cyclohexane/methylene chloride/methanol/formic acid (2:1:1:0.5). Based on TLC analysis, the eluted fractions were combined to yield five fractions, named I–V (I-1.405 g, II-1.77 g, III-7.18 g, IV-3.45 g, V-1.07 g). Fraction III was subjected to column chromatography on silica gel using a gradient solvent system starting with cyclohexane/methylene dichloride 90:10 and successively changing to chloroform/methanol 10:90 providing seven sub-fractions, named A–G (A-0.532 g, B-0.6912 g, C-0.8149 g, D-1.063 g, E-0.1401 g, F-2.8811 g, G-0.2108 g). Sub-fraction E was purified by preparative RP-HPLC (Phenomenex Hyperclone 5 μ) using the mobile phase methanol/water 70:30 containing 0.1% formic acid (flow 8 mL/min). Plakortide E (**2**) (Figure 1) eluted at a retention time of 40 min. The NMR (¹H, ¹³C, 2D NMR) and mass spectrometry (MS) data and the optical rotation were in agreement with those reported previously [31]. Subfraction D was purified by preparative RP-HPLC (Phenomenex Hyperclone 5 μ) using methanol/acetonitrile/water 73:6:21 containing 0.1% formic acid as the mobile phase (flow 9 mL/min). Plakortin (**3**) eluted at a retention time of 18 min. The structure of the compound was analyzed by NMR spectroscopy and mass spectrometry, and according to literature data [28,34,36], the compound was identified as plakortin. Sub-fraction C was purified by semi-preparative RP-HPLC (Phenomenex Hyperclone 5 μ) using acetonitrile/water 60:30 containing 0.1% formic acid as the mobile phase (flow 2 mL/min). Dihydroplakortin (**4**) eluted at a retention time of 41 min. The structure of the compound was elucidated by NMR spectroscopy, mass spectrometry, and optical rotation, and was assigned according to the literature data [37] as dihydroplakortin. The NMR data of the compounds are presented in the Supplementary Materials.

2.3. Semi-Synthesis of Plakortic Acid (5) and Plakortide E Methyl Ester (6)

Plakortin (**3**) was converted into its acid, plakortic acid (**5**), by hydrolysis with LiOH (3 eq.) in THF/water (4:1). After acidic work-up, the residue was further purified via preparative RP-HPLC. The structure of the compound was analyzed by NMR spectroscopy, mass spectrometry, and optical rotation, and, according to the literature data [39], the compound was identified as plakortic acid. Plakortide E (**2**) was converted into its ester (**6**) via Steglich esterification with methanol, DCC, and DMAP. The raw product was further purified via preparative RP-HPLC. The NMR data were in agreement with the literature data [40,41]. The NMR data of the compounds are reported in the Supplementary Materials.

2.4. Cytotoxicity Assay

Drug-sensitive leukemia CCRF-CEM cells and its multi-drug resistant (MDR) subline CEM/ADR5000 were used to test the cytotoxicity of endoperoxides **1–7**. The resazurin reduction assay [43] was performed to determine the cytotoxicity of the seven compounds in a concentration range of 0.001 to 10 µg/mL as previously described [44–48]. Cytotoxicity of established cytostatic drugs against sensitive and multi-drug resistant leukemia cell lines was previously reported by our group (Table 2) [49]. The IC_{50} values were determined from dose response curves and resistance ratios were calculated by dividing the IC_{50} of resistant cells by the IC_{50} of the corresponding parental cells. A degree of resistance >1 indicated that the compound kills the parental cells more effectively than the MDR cells, indicating cross-resistance, while a degree of resistance <1 indicates that the drug kills the MDR cells more effectively, indicating hypersensitivity (collateral sensitivity). The results are shown in Table 2.

Table 2. Cytotoxicity of endoperoxides **1–7** and reference drugs against sensitive and multi-drug resistant leukemia cell lines.

Compound	CCRF-CEM IC_{50} [µM]	CEM/ADR5000 IC_{50} [µM]	Resistance Ratio
6-epi-Plakortide H acid (**1**)	0.18 ± 0.003	0.36 ± 0.01	2.00
Plakortide E (**2**)	1.90 ± 0.09	4.30 ± 0.1	2.26
Plakortin (**3**)	1.97 ± 0.06	2.26 ± 0.08	1.15
Dihydroplakortin (**4**)	1.13 ± 0.11	1.85 ± 0.13	1.64
Plakortic acid (**5**)	0.19 ± 0.004	0.24 ± 0.009	1.26
Plakortide E methyl ester (**6**)	NI [1]	NI [1]	N/A
6-epi-Plakortide H (**7**)	NI [1]	NI [1]	N/A
Doxorubicin *	0.012 ± 0.002	12.2 ± 54.2	1,036
Epirubicin *	0.022 ± 0.003	10.50 ± 3.90	484
Vincristine *	0.002 ± 0.0001	1.04 ± 0.15	613
Docetaxel *	0.0004 ± 0.0001	0.18 ± 0.02	438
Paclitaxel *	0.004 ± 0.0004	0.741 ± 0.137	200

[1] NI, no inhibition at 27 µM; * data taken from reference [49].

3. Discussion

The most obvious structure-activity relationship (SAR) concerns the esters **6**, **7**, and their acid counterparts **2** and **1**: the free acids possessed cytotoxic activity at micromolar concentrations, while the relevant esters were inactive. Similarly, plakortic acid (**5**) was more potent (about 10-fold) than its natural ester plakortin (**3**). Moreover, the side chain did not have any influence on the cytotoxicity (compare **1** and **5**). In contrast, the size of the endoperoxide ring (five-membered vs. six-membered) was important, with the six-membered 6-epi-plakortide H acid (**1**) being 10-fold more active than the five-membered endoperoxide plakortide E (**2**) with the same side chain. Plakortide E (**2**) and its methyl ester (**6**) also possess a double bond activated by an electron-withdrawing substituent (acid or ester) for nucleophilic attack [50], which might also contribute to cytotoxicity. However, the data did not support this assumption, since the methyl ester of plakortide E (**6**) which also contains the activated double bond was inactive.

The inactivity or lower activity of the ester derivatives compared to their acid counterparts was in line with previous findings. For the plakortide H acid and its methyl ester, high cytotoxic effects (IC_{50} <0.7 µg/mL) and inactivity (>100 µg/mL), respectively, were found against the cell lines NIH3T3 (mouse embryo fibroblast), SSVNIH3T3 (Simian sarcoma virus-transformed NIH3T3), and KA3IT (virally transformed NIH3T3) [28]. Cytotoxic activity against tumor cells (including CCRF-CEM) was also reported for the acids plakortide M and N [29]. On the other hand, plakortide F as the methyl ester with a six-membered endoperoxide structure showed some activity against cancer cell lines [51]. Taking into account the facile hydrolysis of methyl esters in vivo but also within cells, the question arises whether the cytotoxic activity of these esters could be attributed at least in part to their acid forms. For the activity in vivo, the methyl esters might be more favourable due to better membrane permeability properties and oral availability compared to the acids. Furthermore, they may act as typical ester pro-drugs.

The degree of resistance of the seven compounds was >1 in all cases, i.e., compounds were more effective against the sensitive cells than against the resistant cells. Plakortic acid (**5**), with comparable IC_{50} values for both cell lines (0.19 µM and 0.24 µM for the sensitive and resistant cells, respectively) seems to be the most promising derivative, since it was highly potent and the resistance ratio was still around 1. However, owing to the fact that CEM/ADR5000 reveal high degrees of cross-resistance (in the range of hundreds to thousands) to standard drugs such as doxorubicin, daunorubicin, vincristine, vinblastine, paclitaxel, docetaxel, and others (Table 2) [49], it is well justified that compounds with degrees of resistance below or around two can be considered as being active against multidrug-resistant cells. In light of better pharmacokinetic properties, the ester derivative plakortin (**3**), which is not as active but displays a similar resistance ratio, may even be the better candidate for further evaluation.

In summary, we present the cytotoxic properties of several plakortide acids and esters. The SAR studies confirmed that the cytotoxic activity is related to the peroxide function as previously shown [52]. In addition, we found that it is also related to the chemical properties of the acid group, versus the ester. Further evaluations will therefore address this question in more detail.

4. Materials and Methods

General Experimental Procedures. Optical rotations were measured with a Krüss Optronic GmbH polarimeter (Hamburg, Germany). ^{1}H spectral data were generated with a Bruker Fourier 300 (300 MHz) and Bruker Avance III 600 (600 MHz, 5 mm TCI-CryoProbe with z-gradient and ATM, SampleXPress Lite 16 sample changer) FT-NMR spectrometer (Karlsruhe, Germany), and the ^{13}C spectral data, COSY, NOESY, DEPT (distortionless enhancement by polarization transfer, HMQC (heteronuclear multiple-quantum correlation), and HMBC (heteronuclear multiple bond correlation) experiments were measured with the 600 MHz Bruker Avance III 600 FT-NMR spectrometer (Karlsruhe, Germany). MS were carried out with a Bruker micrOTOF 88 mass spectrometer (Bremen, Germany) and a LC/MSD-Trap-Mass spectrometer (Agilent Technologies, LC/MSD Ion Trap, Waldbronn, Germany). Column chromatography was performed on silica gel (0.063–0.200 mm mesh, Merck, Darmstadt, Germany). TLC analyses were carried out using pre-coated silica gel 60 F254 plates (0.20 mm, Merck), and spots were visualized using vanillin spray reagent. DCC, DMAP, and reagents were purchased from Sigma-Aldrich (Munich, Germany) or Fluka (Munich, Germany). Solvents were purchased from Roth (Karlsruhe, Germany) or Merck. High performance liquid chromatography was performed on a Varian ProStar analytical/preparative HPLC Linear Upscale system (0.05–50 mL/min at 275 bar pressure with scale-mast), a preparative autosampler and a 2-channel UV detector (Waldbronn, Germany). The detection wavelengths were 254 nm and 230 nm.

4.1. 6-Epi-plakortide H acid (1), [[(3S,4S,6R)-4,6-Diethyl-6-((1E,5E)-4-(R)-ethyl-2-methyl-octa-1,5-dienyl)-[1,2]dioxan-3-yl]-acetic acid]

The methyl ester (7) was hydrolysed using the method described below for plakortic acid (5). The residue was purified using preparative RP-HPLC. Yellow viscous oil (4.1 mg); $[\alpha]_D^{23}$ = −157.84 (*c* 0.0037, CHCl$_3$) (reference [33] reports $[\alpha]_D^{20}$ = −145 (*c* 1.1, CHCl$_3$)); ESI-MS: *m/z* 375.25 [M + Na]$^+$, calcd. for C$_{21}$H$_{36}$O$_4$, 352.51. NMR data are reported in Table 1; since they were found to be identical to those described in reference [33], the compound was identified as the 6-epimer of plakortide H acid.

Plakortide E (2): 18 mg; the $[\alpha]_D^{23}$, ^{1}H and ^{13}C NMR, and MS data were identical in all respects to those previously reported in the literature [31].

Plakortin (3): pale yellow coloured oil (49.8 mg); $[\alpha]_D^{23}$ = +154.93 (*c* 0.0075, CHCl$_3$); [53] (see in the reference $[\alpha]_D^{20}$ = +189 (*c* 2.9, CHCl$_3$)) LC-MS: *m/z* 334.6 [M + Na]$^+$, calcd. for C$_{18}$H$_{32}$O$_4$ *m/z* 312.44; ^{1}H and ^{13}C NMR data were identical in all respects to those previously reported in the literature [28,34,35].

Dihydroplakortin (4): colourless oil (1.8 mg); ESI-MS: *m/z* 337.20 [M + Na]$^+$, calcd. for C$_{18}$H$_{34}$O$_4$ *m/z* 314.46; the optical rotation [53] (see in the reference $[\alpha]_D^{20}$ = +49 (*c* 0.002, CHCl$_3$)) was not determined due to insufficient quantity of the substance. ^{1}H and ^{13}C NMR data were identical in all respects to those previously reported in the literature [37].

Plakortic acid (5): Plakortin (3) was converted into its acid, plakortic acid, by hydrolysis. To a solution of plakortin (43.2 mg, 0.138 mmol) in THF/H$_2$O (4:1; 10 mL), LiOH (17.4 mg, 3 eq.) was added at 0 °C. The reaction mixture was allowed to warm to room temperature and allowed to stir for 24 h. The reaction was monitored using TLC until the starting material disappeared. Then the reaction mixture was acidified to pH 2 with 10% aqueous HCl and extracted with ether (3 × 10 mL). The combined extracts were washed with NaCl solution (15 mL) and dried over anhydrous Na$_2$SO$_4$. The residue was further purified via preparative RP-HPLC. Colourless oil (4.1 mg), $[\alpha]_D^{23}$ = +109 (*c* 0.002, CHCl$_3$); LC-MS: *m/z* 321.2 [M + Na]$^+$, calcd. for C$_{17}$H$_{30}$O$_4$ *m/z* 298.42. ^{1}H and ^{13}C NMR data were identical in all respects to those previously reported in the literature [39].

Plakortide E methyl ester (6): Plakortide E (2) was converted into its ester form via Steglich esterification. To a solution of plakortide E (9.6 mg, 0.0274 mmol in dichloromethane at 0 °C), methanol (0.88 mL, 0.4314 mmol, 1.0 eq.) was first added; then, 1.05 eq. DCC (6.01 mg, 0.0291 mmol) and 0.1 eq. DMAP (0.5 mg, 0.0041 mmol) were added. The reaction mixture was stirred for 1 h at 0 °C and then at room temperature for 24 h. The colourless solid by-product *N,N'*-dicyclohexylurea was filtered off and the organic phase was washed with half-saturated solutions of ammonium chloride, sodium bicarbonate, and sodium chloride. It was then dried over sodium sulphate, filtered off, and the organic phase was removed in vacuo. The raw product was further purified via preparative RP-HPLC (Phenomenex Hyperclone 5 μ) using methanol/acetonitrile/water 85:6:9 containing 0.1% formic acid (flow 9 mL/min). Plakortide E methyl ester eluted at 14 min. Colourless viscous oil (3.2 mg, 33%); $[\alpha]_D^{23}$ = +74.1 (*c* 0.00305, CHCl$_3$); LC-MS: *m/z* 403.9 [M + K]$^+$, calcd. for C$_{22}$H$_{36}$O$_4$ *m/z* 364.52. ^{1}H and ^{13}C NMR data were identical in all respects to those previously reported in the literature [40,41].

6-Epi-Plakortide H (7): The fraction containing several acids (1mix) was converted into an ester mixture (7mix) using the Steglich esterification procedure as described above for plakortide E methyl ester (6). Then, the mixture was purified using preparative RP-HPLC to yield the pure ester (7) which eluted at 14 min. Pale yellow viscous oil (6.6 mg), $[\alpha]_D^{23}$ = −107.14 (*c* 0.0028, CHCl$_3$) [53] (see in the reference plakortide H methyl ester, $[\alpha]_D^{20}$ = +5.5 (*c* 2.9, CHCl$_3$), 4-epi-plakortide H methyl ester $[\alpha]_D^{20}$ = +19 (*c* 0.13, CHCl$_3$)). LC-MS: *m/z* 389.1 [M + Na]$^+$, calcd. for C$_{22}$H$_{38}$O$_4$ *m/z* 366.53. The absolute configuration was assigned as 6R, 10R in analogy with that of the 6-epi-plakortide H acid (1).

4.2. Cytotoxicity Assays

The origin and the maintenance of the cell lines were reported previously [45–47]. The resazurin reduction assay [43] was performed to determine the cytotoxicity of the seven compounds in a concentration range of 0.001 to 10 µg/mL as previously described [47,48].

Supplementary Materials: The following are available online at www.mdpi.com/1660-3397/15/3/63/s1, Table S1: NMR data of plakortin (**3**); Table S2: Dihydroplakortin (**4**); Table S3: Plakortic acid (**5**); Table S4: Plakortide E methyl ester (**6**); Table S5: 6-epi-Plakortide H (methyl ester) (**7**); Figure S1: Structures of natural (**1**, **2**, **3**, **4**) and semi-synthetic (**5**, **6**, **7**) endoperoxides from a sample of the sponge *Plakortis halichondrioides*: 6-epi-plakortide H acid (**1**), plakortin E acid (**2**), plakortin (**3**), dihydroplakortin (**4**), plakortic acid (**5**), plakortide E methyl ester (**6**), and 6-epi-plakortide H (**7**).

Acknowledgments: This work was partially funded by the European Union's Seventh Framework Programme (FP7) 2007–2013 under Grant Agreement No. 311848 (Bluegenics). The sponge was collected during an oceanographic expedition managed by J. R. Pawlik, University of North Carolina at Wilmington that we wish to thank for helping in collection and identification of the sample. Sponge collection was made possible by UNOLS (University-National Oceanographic Laboratory System) funding through a grant from the US-NSF (US National Science Foundation) Biological Oceanography Program (OCE 1029515) and the crew of the R/V Walton Smith (University of Miami). Sponge collection was made possible under Permit MAF/LIA/22 from the Department of Marine Resources of the Bahamas. The authors thank Ute Hentschel-Humeida, Marine Microbiology, GEOMAR, Helmholtz Centre for Ocean Research, Kiel, Germany, for critical discussions.

Author Contributions: Tanja Schirmeister and Swarna Oli conceived and designed the experiments concerning sponge extraction, isolation of compounds, and semi-synthesis; Swarna Oli and Hongmei Wu performed these experiments; Tanja Schirmeister, Swarna Oli and Hongmei Wu analyzed the NMR data; Tanja Schirmeister, Thomas Efferth, Gerardo Della Sala and Valeria Costantino wrote the paper.

Conflicts of Interest: The authors declare no conflict of interest. The founding sponsors had no role in the design of the study; in the collection, analyses, or interpretation of data; in the writing of the manuscript, and in the decision to publish the results.

Abbreviations

DCC	dicyclohexylcarbodiimide
DMAP	4-dimethylamino pyridine
MDR	multi-drug resistant
NMR	nuclear magnetic resonance
RP-HPLC	reversed phase high performance liquid chromatography
SAR	structure-activity relationship
TLC	thin layer chromatography

References

1. Lucas, R.; Casapullo, A.; Ciasullo, L.; Gomez Paloma, L.; Payà, M. Cycloamphilectenes, a new type of potent marine diterpenes: Inhibition of nitric oxide production in murine macrophages. *Life Sci.* **2003**, *72*, 2543–2552. [CrossRef]

2. Costantino, V.; Fattorusso, E.; Mangoni, A.; Perinu, C.; Cirino, G.; De Gruttola, L.; Roviezzo, F. Tedanol: A potent anti-inflammatory ent-pimarane diterpene from the Caribbean sponge *Tedania ignis*. *Bioorg. Med. Chem.* **2009**, *17*, 7542–7547. [CrossRef] [PubMed]

3. Teta, R.; Irollo, E.; Della Sala, G.; Pirozzi, G.; Mangoni, A.; Costantino, V. Smenamides A and B, chlorinated peptide/polyketide hybrids containing a dolapyrrolidinone unit from the Caribbean sponge *Smenospongia aurea*. Evaluation of their role as leads in antitumor drug research. *Mar. Drugs* **2013**, *11*, 4451–4463. [CrossRef] [PubMed]

4. Esposito, G.; Miceli, R.; Ceccarelli, L.; Della Sala, G.; Irollo, E.; Mangoni, A.; Teta, R.; Pirozzi, G.; Costantino, V. Isolation and assessing the anti-proliferative activity in vitro of smenothiazole A and B, chlorinated thiazole-containing peptide/polyketides from the Caribbean sponge *Smenospongia aurea*. *Mar. Drugs* **2015**, *13*, 444–459. [CrossRef] [PubMed]

5. Costantino, V.; Fattorusso, E.; Imperatore, C.; Mangoni, A. Glycolipids from sponges. 20. *J*-coupling analysis for stereochemical assignments in furanosides: Structure elucidation of vesparioside B, a glycosphingolipid from the marine sponge *Spheciospongia vesparia*. *J. Org. Chem.* **2008**, *73*, 6158–6165. [CrossRef] [PubMed]

6. Costantino, V.; Fattorusso, E.; Imperatore, C.; Mangoni, A. Ectyoceramide, the first natural hexofuranosylceramide from the marine sponge *Ectyoplasia ferox. Eur. J. Org. Chem.* **2003**, *2003*, 1433–1437. [CrossRef]
7. Blunt, J.W.; Copp, B.R.; Keyzers, R.A.; Munro, M.H.G.; Prinsep, M.R. Marine natural products. *Nat. Prod. Rep.* **2016**, *33*, 382–431. [CrossRef] [PubMed]
8. Efferth, T.; Zacchino, S.; Georgiev, M.I.; Liu, L.; Wagner, H.; Panossian, A. Nobel Prize for artemisinin brings phytotherapy into the spotlight. *Phytomedicine* **2015**, *22*, A1–A3. [CrossRef] [PubMed]
9. Efferth, T.; Rücker, G.; Falkenberg, M.; Manns, D.; Olbrich, A.; Fabry, U.; Osieka, R. Detection of apoptosis in KG-1a leukemic cells treated with investigational drugs. *Arzneimittelforschung* **1996**, *46*, 196–200. [PubMed]
10. Efferth, T.; Dunstan, H.; Sauerbrey, A.; Miyachi, H.; Chitambar, C.R. The anti-malarial artesunate is also active against cancer. *Int. J. Oncol.* **2001**, *18*, 767–773. [CrossRef] [PubMed]
11. Berger, T.G.; Dieckmann, D.; Efferth, T.; Schultz, E.S.; Funk, J.O.; Baur, A.; Schuler, G. Artesunate in the treatment of metastatic uveal melanoma—First experiences. *Oncol. Rep.* **2005**, *14*, 1599–1603. [CrossRef] [PubMed]
12. Sieber, S.; Gdynia, G.; Roth, W.; Bonavida, B.; Efferth, T. Combination treatment of malignant B-cells using the anti-CD20 antibody rituximab and the anti-malarial artesunate. *Int. J. Oncol.* **2009**, *35*, 149–158. [PubMed]
13. Jansen, F.H.; Adoubi, I.; Kouassi Comoe, J.C.; DE Cnodder, T.; Jansen, N.; Tschulakow, A.; Efferth, T. First study of oral artenimol-R in advanced cervical cancer: Clinical benefit, tolerability and tumor markers. *Anticancer Res.* **2011**, *31*, 4417–4422. [PubMed]
14. Rutteman, G.R.; Erich, S.A.; Mol, J.A.; Spee, B.; Grinwis, G.C.; Fleckenstein, L.; London, C.A.; Efferth, T. Safety and efficacy field study of artesunate for dogs with non-resectable tumours. *Anticancer Res.* **2013**, *33*, 1819–1827. [PubMed]
15. Krishna, S.; Ganapathi, S.; Ster, I.C.; Saeed, M.E.; Cowan, M.; Finlayson, C.; Kovacsevics, H.; Jansen, H.; Kremsner, P.G.; Efferth, T.; et al. A randomised, double blind, placebo-controlled pilot study of oral artesunate therapy for colorectal cancer. *EBioMedicine* **2014**, *2*, 82–90. [CrossRef] [PubMed]
16. Efferth, T.; Olbrich, A.; Sauerbrey, A.; Ross, D.D.; Gebhart, E.; Neugebauer, M. Activity of ascaridol from the anthelmintic herb *Chenopodium anthelminticum* L. against sensitive and multidrug-resistant tumor cells. *Anticancer Res.* **2002**, *22*, 4221–4224. [PubMed]
17. Abbasi, R.; Efferth, T.; Kuhmann, C.; Opatz, T.; Hao, X.; Popanda, O.; Schmezer, P. The endoperoxide ascaridol shows strong differential cytotoxicity in nucleotide excision repair-deficient cells. *Toxicol. Appl. Pharmacol.* **2012**, *259*, 302–310. [CrossRef] [PubMed]
18. Varoglu, M.; Peters, B.M.; Crews, P. The structures and cytotoxic properties of polyketide peroxides from a *Plakortis* sponge. *J. Nat. Prod.* **1995**, *58*, 27–36. [CrossRef] [PubMed]
19. Valeriote, F.A.; Tenney, K.; Media, J.; Pietraszkiewicz, H.; Edelstein, M.; Johnson, T.A.; Amagata, T.; Crews, P. Discovery and development of anticancer agents from marine sponges: Perspectives based on a chemistry-experimental therapeutics collaborative program. *J. Exp. Ther. Oncol.* **2012**, *10*, 119–134. [PubMed]
20. Rubush, D.M.; Morges, M.A.; Rose, B.J.; Thamm, D.H.; Rovis, T. An asymmetric synthesis of 1,2,4-trioxane anticancer agents via desymmetrization of peroxyquinols through a brønsted acid catalysis cascade. *J. Am. Chem. Soc.* **2012**, *134*, 13554–13557. [CrossRef] [PubMed]
21. Opsenica, D.; Angelovski, G.; Pocsfalvi, G.; Juranić, Z.; Žižak, Ž.; Kyle, D.; Milhous, W.K.; Šolaja, B.A. Antimalarial and antiproliferative evaluation of bis-steroidal tetraoxanes. *Bioorg. Med. Chem.* **2003**, *11*, 2761–2768. [CrossRef]
22. Parrish, J.D.; Ischay, M.A.; Lu, Z.; Guo, S.; Peters, N.R.; Yoon, T.P. Endoperoxide synthesis by photocatalytic aerobic [2+2+2] cycloadditions. *Org. Lett.* **2012**, *14*, 1640–1643. [CrossRef]
23. Van Huijsduijnen, R.H.; Guy, R.K.; Chibale, K.; Haynes, R.K.; Peitz, I.; Kelter, G.; Phillips, M.A.; Vennerstrom, J.L.; Yuthavong, Y.; Wells, T.N.C. Anticancer properties of distinct antimalarial drug classes. *PLoS ONE* **2013**, *8*, e82962. [CrossRef] [PubMed]
24. Yaremenko, I.A.; Syroeshkin, M.A.; Levitsky, D.O.; Fleury, F.; Terent'ev, A.O. Cyclic peroxides as promising anticancer agents: In Vitro cytotoxicity study of synthetic ozonides and tetraoxanes on human prostate cancer cell lines. *Med. Chem. Res.* **2017**, *26*, 170–179. [CrossRef]
25. Terzić, N.; Opsenica, D.; Milić, D.; Tinant, B.; Smith, K.S.; Milhous, W.K.; Šolaja, B.A. Deoxycholic acid-derived tetraoxane antimalarials and antiproliferatives. *J. Med. Chem.* **2007**, *50*, 5118–5127. [CrossRef] [PubMed]

26. Norris, M.D.; Perkins, M.V. Structural diversity and chemical synthesis of peroxide and peroxide-derived polyketide metabolites from marine sponges. *Nat. Prod. Rep.* **2016**, *33*, 861–880. [CrossRef] [PubMed]

27. Della Sala, G.; Hochmuth, T.; Teta, R.; Costantino, V.; Mangoni, A. Polyketide synthases in the microbiome of the marine sponge *Plakortis halichondrioides*: A metagenomic update. *Mar. Drugs* **2014**, *12*, 5425–5440. [CrossRef] [PubMed]

28. Hoye, T.R.; Alarif, W.M.; Basaif, S.S.; Abo-Elkarm, M.; Hamann, M.T.; Wahba, A.E.; Ayyad, S.N. New cytotoxic cyclic peroxide acids from *Plakortis* sp. marine sponge. *ARKIVOC* **2015**, *2015*, 164–175. [PubMed]

29. Jimenez, M.D.; Garzon, S.P.; Rodriguez, A.D. Plakortides M and N, bioactive polyketide endoperoxides from the Caribbean marine sponge *Plakortis halichondrioides*. *J. Nat. Prod.* **2003**, *66*, 655–661. [CrossRef] [PubMed]

30. Rudi, A.; Kashman, Y. Three new cytotoxic metabolites from the marine sponge *Plakortis halichondrioides*. *J. Nat. Prod.* **1993**, *56*, 1827–1830. [CrossRef] [PubMed]

31. Oli, S.; Abdelmohsen, U.R.; Hentschel, U.; Schirmeister, T. Identification of plakortide E from the Caribbean sponge *Plakortis halichondroides* as a trypanocidal protease inhibitor using bioactivity-guided fractionation. *Mar. Drugs* **2014**, *12*, 2614–2622. [CrossRef] [PubMed]

32. Patil, A.D.; Freyer, A.J.; Carte, B.; Johnson, R.K.; Lahouratate, P. Plakortides, novel cyclic peroxides from the sponge *Plakortis halichondrioides*: activators of cardiac SR-Ca^{2+}-pumping ATPase. *J. Nat. Prod.* **1996**, *59*, 219–223. [CrossRef] [PubMed]

33. Santos, E.A.; Quintela, A.L.; Ferreira, E.G.; Sousa, T.S.; Pinto, F.D.C.; Hajdu, E.; Carvalho, M.S.; Salani, S.; Rocha, D.D.; Wilke, D.V.; et al. Cytotoxic plakortides from the Brazilian marine sponge *Plakortis angulospiculatus*. *J. Nat. Prod.* **2015**, *78*, 996–1004. [CrossRef] [PubMed]

34. Higgs, M.D.; Faulkner, D.J. Plakortin, an antibiotic from *Plakortis halichondrioides*. *J. Org. Chem.* **1978**, *43*, 3454–3457. [CrossRef]

35. Kossuga, M.H.; Nascimento, A.M.; Reimão, J.Q.; Tempone, A.G.; Taniwaki, N.N.; Veloso, K.; Ferreira, A.G.; Cavalcanti, B.C.; Pessoa, C.; Moraes, M.O.; et al. Antiparasitic, antineuroinflammatory, and cytotoxic polyketides from the marine sponge *Plakortis angulospiculatus* collected in Brazil. *J. Nat. Prod.* **2008**, *71*, 334–339. [CrossRef] [PubMed]

36. Chianese, G.; Persico, M.; Yang, F.; Lin, H.-W.; Guo, Y.W.; Basilico, N.; Parapini, S.; Taramelli, D.; Taglialatela-Scafati, O.; Fattorusso, C. Endoperoxide polyketides from a Chinese *Plakortis simplex*: Further evidence of the impact of stereochemistry on antimalarial activity of simple 1,2-dioxanes. *Bioorg. Med. Chem.* **2014**, *22*, 4572–4580. [CrossRef] [PubMed]

37. Fattorusso, E.; Parapini, S.; Campagnuolo, C.; Basilico, N.; Taglialatela-Scafati, O.; Taramelli, D. Activity against *Plasmodium falciparum* of cycloperoxide compounds obtained from the sponge *Plakortis simplex*. *J. Antimicrob. Chemother.* **2002**, *50*, 883–888. [CrossRef] [PubMed]

38. Blunt, J.W.; Munro, M.H.G. (Eds.) *Dictionary of Marine Natural Products*; Chapman & Hall/CRC: London, UK, 2015.

39. Phillipson, D.W.; Rinehart, K.L. Antifungal peroxide-containing acids from two Caribbean sponges. *J. Am. Chem. Soc.* **1983**, *105*, 7735–7736. [CrossRef]

40. Patil, A.D.; Freyer, A.J.; Bean, M.F.; Carte, B.K.; Westley, J.W.; Johnson, R.K.; Lahouratate, P. The plakortones, novel bicyclic lactones from the sponge *Plakortis halichondrioides*: Activators of cardiac SR-Ca^{2+}-pumping ATPase. *Tetrahedron* **1996**, *52*, 377–394. [CrossRef]

41. Sun, X.Y.; Tian, X.Y.; Li, Z.W.; Peng, X.S.; Wong, H.N. Total synthesis of plakortide E and biomimetic synthesis of plakortone B. *Chemistry* **2011**, *17*, 5874–5880. [CrossRef] [PubMed]

42. The Sponge Guide. Available online: http://www.spongeguide.org (accessed on 7 July 2013).

43. O'Brien, J.; Wilson, I.; Orton, T.; Pognan, F. Investigation of the Alamar blue (resazurin) fluorescent dye for the assessment of mammalian cell cytotoxicity. *Eur. J. Biochem.* **2000**, *267*, 5421–5426. [CrossRef] [PubMed]

44. Kimmig, A.; Gekeler, V.; Neumann, M.; Frese, G.; Handgretinger, R.; Kardos, G.; Diddens, H.; Niethammer, D. Susceptibility of multidrug-resistant human leukemia cell lines to human interleukin 2-activated killer cells. *Cancer Res.* **1990**, *50*, 6793–6799. [PubMed]

45. Brugger, D.; Herbart, H.; Gekeler, V.; Seitz, G.; Liu, C.; Klingebiel, T.; Orlikowsky, T.; Einsele, H.; Denzlinger, C.; Bader, P.; et al. Functional analysis of P-glycoprotein and multidrug resistance associated protein related multidrug resistance in AML-blasts. *Leuk. Res.* **1999**, *23*, 467–475. [CrossRef]

46. Efferth, T.; Sauerbrey, A.; Olbrich, A.; Gebhart, E.; Rauch, P.; Weber, H.O.; Hengstler, J.G.; Halatsch, M.E.; Volm, M.; Tew, K.D.; et al. Molecular modes of action of artesunate in tumor cell lines. *Mol. Pharmacol.* **2003**, *64*, 382–394. [CrossRef] [PubMed]

47. Kuete, V.; Ngameni, B.; Wiench, B.; Krusche, B.; Horwedel, C.; Ngadjui, B.T.; Efferth, T. Cytotoxicity and mode of action of four naturally occurring flavonoids from the genus *Dorstenia*: Gancaonin Q, 4-hydroxylonchocarpin, 6-prenylapigenin, and 6,8-diprenyleriodictyol. *Planta Med.* **2011**, *77*, 1984–1989. [CrossRef] [PubMed]

48. Ooko, E.; Alsalim, T.; Saeed, B.; Saeed, M.E.; Kadioglu, O.; Abbo, H.S.; Titinchi, S.J.; Efferth, T. Modulation of P-glycoprotein activity by novel synthetic curcumin derivatives in sensitive and multidrug-resistant T-cell acute lymphoblastic leukemia cell lines. *Toxicol. Appl. Pharmacol.* **2016**, *305*, 216–233. [CrossRef] [PubMed]

49. Efferth, T.; Konkimalla, V.B.; Wang, Y.F.; Sauerbrey, A.; Meinhardt, S.; Zintl, F.; Mattern, J.; Volm, M. Prediction of broad spectrum resistance of tumors towards anticancer drugs. *Clin. Cancer Res.* **2008**, *14*, 2405–2412. [CrossRef] [PubMed]

50. Schultz, T.W.; Yarbrough, J.W. Trends in structure-toxicity relationships for carbonyl-containing α,β-unsaturated compounds. *SAR QSAR Environ. Res.* **2004**, *15*, 139–146. [CrossRef] [PubMed]

51. Gochfeld, D.J.; Hamann, M.T. Isolation and biological evaluation of filiformin, plakortide F, and plakortone G from the Caribbean sponge *Plakortis* sp. *J. Nat. Prod.* **2001**, *64*, 1477–1479. [CrossRef] [PubMed]

52. Fattorusso, C.; Campiani, G.; Catalanotti, B.; Persico, M.; Basilico, N.; Parapini, S.; Taramelli, D.; Campagnuolo, C.; Fattorusso, E.; Romano, A.; et al. Endoperoxide derivatives from marine organisms: 1,2-dioxanes of the plakortin family as novel antimalarial agents. *J. Med. Chem.* **2006**, *49*, 7088–7094. [CrossRef] [PubMed]

53. Blunt, J.W.; Munro, M.H.G. (Eds.) *Dictionary of Marine Natural Products*; Chapman & Hall/CRC: New York, NY, USA, 2008.

marine drugs

MDPI

Article

Brevianamides and Mycophenolic Acid Derivatives from the Deep-Sea-Derived Fungus *Penicillium brevicompactum* DFFSCS025

Xinya Xu, Xiaoyong Zhang, Xuhua Nong, Jie Wang and Shuhua Qi *

Key Laboratory of Tropical Marine Bio-resources and Ecology, Guangdong Key Laboratory of Marine Materia Medica, RNAM Center for Marine Microbiology, South China Sea Institute of Oceanology, Chinese Academy of Sciences, 164 West Xingang Road, Guangzhou 510301, China; xuxinya@scsio.ac.cn (X.X.); zhangxiaoyong@scsio.ac.cn (X.Z.); xhnong@scsio.ac.cn (X.N.); wangjielangjing@126.com (J.W.)
* Correspondence: shuhuaqi@scsio.ac.cn; Tel.: +86-20-8902-2112; Fax: +86-20-8445-8964

Academic Editors: Sergey A. Dyshlovoy and Friedemann Honecker
Received: 29 December 2016; Accepted: 10 February 2017; Published: 17 February 2017

Abstract: Four new compounds (**1–4**), including two brevianamides and two mycochromenic acid derivatives along with six known compounds were isolated from the deep-sea-derived fungus *Penicillium brevicompactum* DFFSCS025. Their structures were elucidated by spectroscopic analysis. Moreover, the absolute configurations of **1** and **2** were determined by quantum chemical calculations of the electronic circular dichroism (ECD) spectra. Compound **9** showed moderate cytotoxicity against human colon cancer HCT116 cell line with IC_{50} value of 15.6 μM. In addition, **3** and **5** had significant antifouling activity against *Bugula neritina* larval settlement with EC_{50} values of 13.7 and 22.6 μM, respectively. The NMR data of **6**, **8**, and **9** were assigned for the first time.

Keywords: *Penicillium brevicompactum*; Brevianamide; Mycochromenic acid derivative; cytotoxicity; antifouling

1. Introduction

Deep-sea-derived microorganisms are new potential resources for discovery of bioactive secondary metabolites [1–3]. In our ongoing search for active compounds from marine fungi, four brevianamides and five mycochromenic acid derivatives were obtained from the deep-sea-derived fungus *Penicillium brevicompactum* DFFSCS025. Brevianamides, a class of indole alkaloids, were isolated from *P. brevicompactum* in 1969 for the first time [4]. Their unique bicyclo[2.2.2]diazaoctane skeleton and multiple bio-activities were attractive to scientists. Most of them exhibited anti-bacterial, anti-insect pests and antitubercular potentials [5,6]. Several brevianamides have been totally synthesized [7–9]. Mycophenolic acid, a phenyl derivative, was found from *Penicillium* sp. in 1893 for the first time [10]. It was an inhibitor of human inosine 5'-monophosphate dehydrogenase (IMPDH), a target for immunosuppressive chemotherapy [11]. Mycophenolic acid and its derivative mycophenolate mofetil have been used as immunosuppressant drugs in the management of auto-immune disorders since the 1990s [12]. Because of instrument limitations, some brevianamides and mycophenolate acid derivatives are short of reliable spectral data including NMR data [4,10,13]. Herein, we describe the separation, structure elucidation, and bioactivities of Compounds **1–10** (Figure 1). The NMR data of **6**, **8**, and **9** were assigned for the first time.

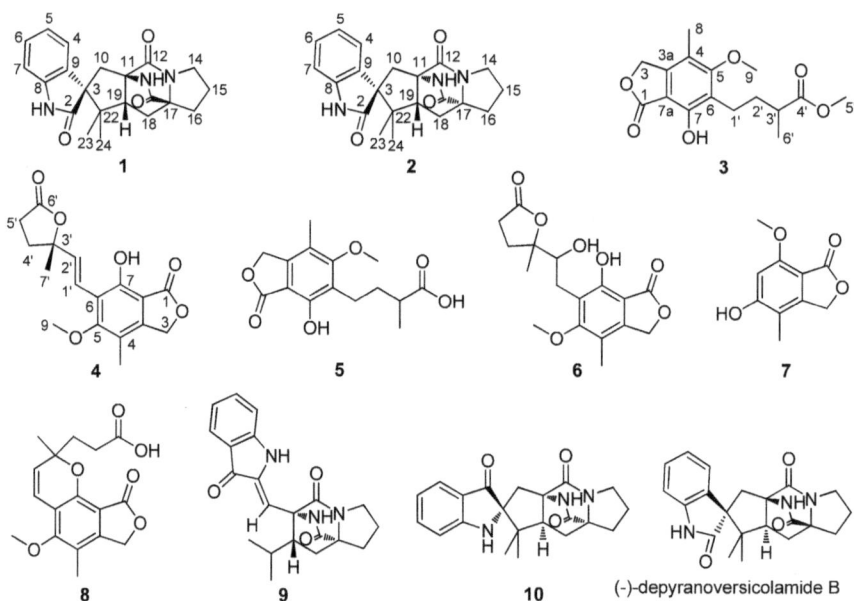

Figure 1. Structures of Compounds **1–10**.

2. Results and Discussion

The deep-sea-derived fungal stain DFFSCS025 was inoculated in liquid medium and fermented in standing situation for 32 days at 28 °C. The culture broths were absorbed with XAD-16 resin; meanwhile, mycelium portions were extracted with 80% acetone. The combined extract (24 g from 30 L) was subjected to silica gel column, ODS column, Sephadex LH-20, and purified with semi-preparation HPLC to yield Compounds **1–10**.

Brevianamide X (**1**) was obtained as yellowish powder. Its molecular formula of $C_{21}H_{23}N_3O_3$ was established by HRESIMS (*m/z* 366.1810 [M + H]$^+$). The ^{1}H NMR spectrum (Table 1) revealed the presence of two methyls at δ_H 0.72 (3H, s) and 0.74 (3H, s), four aromatic protons at δ_H 6.83 (1H, d, *J* = 7.6 Hz), 6.98 (1H, td, *J* = 0.8, 7.6, 7.6 Hz), 7.20 (1H, td, *J* = 0.9, 7.6, 7.6 Hz), 7.23 (1H, d, *J* = 7.5 Hz), and two exchangeable protons at δ_H 9.13 (1H, br s) and 10.36 (1H, br s). The ^{13}C NMR spectrum (Table 1) exhibited 21 carbons including two methyls (δ_C 20.2, 23.7), five methylenes (δ_C 24.8, 29.5, 29.9, 33.6, 43.8), five methines (δ_C 55.9, 109.6, 121.5, 126.4, 128.3), and nine quaternary carbons (δ_C 45.6, 61.9, 66.1, 68.5, 131.0, 142.8, 169.8, 173.5, 182.8). These NMR data were similar to those of (−)-depyranoversicolamide B (**11**) [7] except the little differences of the chemical shifts of C-3/11/19/20/22. Detailed analysis of 2D NMR spectra revealed that **1** had the same planar structure as (−)-depyranoversicolamide B (**11**) (Figure 2). The relative configuration of **1** was further determined by NOESY spectrum. The NOE correlations between H-10β, H-19, and H-21 established that they were on the same side, while NOE correlation between H-4 and H-10α indicated that they were on the other side (Figure 3). The relative configuration of **1** was therefore proposed to be 3S*, 11R*, 17R*, and 19R*. In order to assign the absolute configuration of **1**, we carried out molecular mechanics calculation using DFT method at B3LYP/6-31G (d) level [14,15]. Furthermore, ECD/TDDFT calculations of all low-energy conformer afforded ECD spectra consistent with the experimental spectrum (Figure 4). The results indicated the 3S, 11R, 17R, 19R configuration for **1** on the basis of the relative configuration. Therefore, Compound **1** was inferred to be a diastereomer of (−)-depyranoversicolamide B.

Table 1. ^{1}H NMR data (500 MHz) and ^{13}C NMR data (125 MHz) of **1** and **2** in DMSO-d_6.

No.	1		2	
	δ_C	δ_H	δ_C	δ_H
1-NH	-	10.36 s	-	10.31 s
2	182.8 C		182.3 C	
3	61.9 C		62.6 C	
4	126.4 CH	7.23, d (7.5)	126.8 CH	7.43, d (7.5)
5	121.5 CH	6.98, dd (7.5, 7.6)	121.3 CH	6.99, dd (7.5, 7.6)
6	128.3 CH	7.20, dd (7.6, 7.6)	128.6 CH	7.20, dd (7.6, 7.6)
7	109.6 CH	6.83, d (7.6)	109.5 CH	6.81, d (7.6)
8	142.8 C		142.9 C	
9	131.0 C		130.2 C	
10	33.6 CH$_2$	2.20, d (14.2)	34.1 CH$_2$	2.14, d (15.2)
		2.86, d (14.2)		2.83, d (15.2)
11	66.1 C		67.6 C	
12	169.8 C		169.5 C	
14	43.8 CH$_2$	3.41, m	43.7 CH$_2$	3.30, m
15	24.8 CH$_2$	1.80, m	24.9 CH$_2$	1.83, m
		1.99, m		2.00, dd (5.9, 12.1)
16	29.5 CH$_2$	2.50, overlapped	28.9 CH$_2$	1.79, m
				2.47, dd (6.4, 12.1)
17	68.5 C		69.1 C	
18	29.9 CH$_2$	1.78, dd (8.2, 12.8)	28.4 CH$_2$	1.79, m
		1.93, dd (10.4, 12.9)		2.47, dd (6.4, 12.1)
19	55.9 CH	3.23, dd (8.3, 10.1)	50.5 CH	3.18, dd (5.0, 10.0)
20	173.5 C		173.0 C	
21-NH	-	9.13, s	-	8.81, s
22	45.6 C		47.3 C	
23	20.2 CH$_3$	0.74, s	20.9 CH$_3$	1.00, s
24	23.7 CH$_3$	0.72, s	23.5 CH$_3$	0.69, s

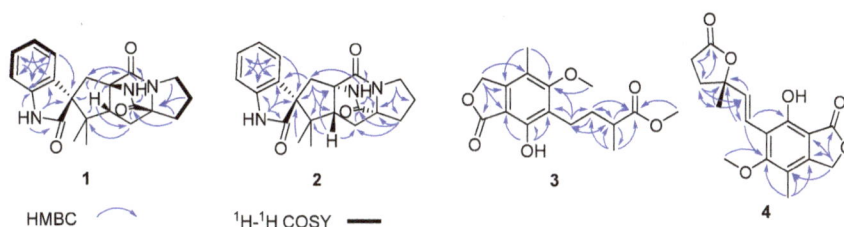

HMBC ⌒→ ^{1}H-^{1}H COSY ——

Figure 2. Key HMBC and COSY correlations of **1–4**.

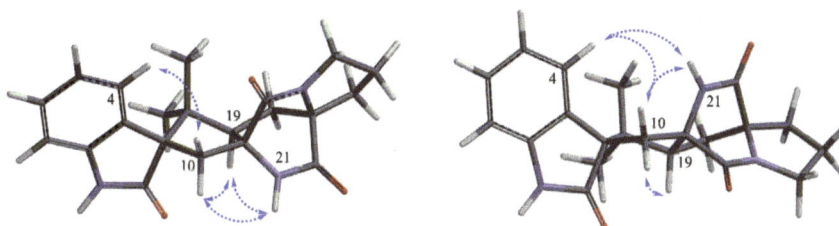

Figure 3. Key NOESY correlations (dashed arrows) of **1** (left) and **2** (right).

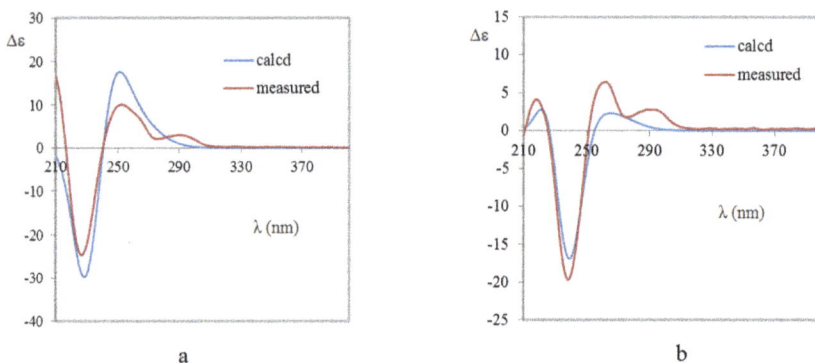

Figure 4. Comparison of the measured and calculated ECD spectra of **1** (a) and **2** (b). (a) ECD spectra of (3*S*, 11*R*, 17*R*, 19*R*)-**1** in MeOH (σ = 0.3 eV, shift = −3 nm); (b) ECD spectra of (3*S*, 11*S*, 17*S*, 19*R*)-**2** in MeOH (σ = 0.27 eV, shift = −2 nm).

Brevianamide Y (**2**) had same molecular formula of $C_{21}H_{23}N_3O_3$ as **1** according to HRESIMS (*m/z* 366.1813 [M + H]$^+$). The NMR data of **2** showed great similarity to those of **1** with the only obvious difference of the high-field shift of C-19 (from δ_C 55.9 in **1** to δ_C 50.5 in **2**) (Table 1). Detailed analysis of 2D NMR spectra revealed that **2** had the same planar structure as **1** (Figure 2). In the NOESY spectrum, NOE correlations between H-4, H-10α, H-21, and H-23 suggested they were on the same side, while NOE correlation between H-19 and H-24 disclosed that they were on the other side (Figure 3). The relative configuration of **2** was suggested as 3*S**, 11*S**, 17*S**, and 19*R**. The absolute configuration of **2** was also determined by molecular mechanics calculation and quantum chemical computations [14,15]. The calculated ECD curve of 3*S*, 11*S*, 17*S*, and 19*R* was consistent with experimental one (Figure 4). The structure of **2** was confirmed as (3*S*, 11*S*, 17*S*, 19*R*)-brevianamide Y.

The molecular formula of **3** was determined as $C_{16}H_{20}O_6$ by the HRESIMS (*m/z* 331.1164 [M + Na]$^+$). The NMR data (Table 2) of **3** was closely resembled with known compound 6-(3-carboxybutyl)-7-hydroxy-5-methoxy-4-methylphthalan-1-one (**5**) [10] except for the presence of an extra methoxy group (δ_H 3.68 and δ_C 51.6). The HMBC correlations from H$_3$-5' to C-4' suggested that the methoxy group was linked to the C-4' of mycophenolic acid skeleton (Figure 2). No specific optical rotation ([α]$_D^{25}$ 0 (*c* 0.1, MeOH)) or circular dichroism spectral data (±0) indicated **3** was isolated as a racemic mixture. We failed to separate the racemic mixture using chiral column by HPLC (CHIRALPAK® IC, Daicel Corporation, Osaka, Japan). Compound **3** was named as 6-(methyl 3-methylbutanoate)-7-hydroxy-5 -methoxy-4-methylphthalan-1-one.

The molecular formula of **4** was determined as $C_{17}H_{18}O_6$ by HRESIMS (*m/z* 341.1005 [M + Na]$^+$). The similar NMR data (Table 2) of **4** and **3** indicated they had the same benzofuranone skeleton. The down-field shift of C-6 (from δ_C 122.5 in **3** to δ_C 117.9 in **4**) was induced by the hyperconjugation of a double bond (C-1', δ_C 118.23; C-2', δ_C 137.38) with benzofuranone skeleton, which was confirmed by the HMBC correlations between H-1'/H-2' and C-6 (Figure 3). The *E*-configured double bond $\Delta^{1'2'}$ was determined by the coupling constant value of approximately 16 Hz between H-1' and H-2'. The NMR data (δ_H 2.19, 2.24, 2.56, 2.64; δ_C 26.4, 28.8, 34.0, 85.8, 177.0) and HMBC correlations from H-7' to C-3', C-4', from H-4 to C-3', C-5', C-6', C-7', and H-5' to C-4', C-6' elucidated the presence of a methyl-furanone residue [16,17] attached at C-2' according to the HMBC correlations from H-1' to C-3', and from H-2' to C-3'. Compound **4** had one chiral center at C-3'. The configuration of **4** was determined by molecular mechanics calculation and quantum chemical computations [14,15]. The (3'*S*)-**4** of calculated ECD curve was consistent with experimental one (Figure 5). Compound **4** was named as (3'*S*)-(*E*)-7-hydroxy-5-methoxy-4-methyl-6-(2-(2-methyl-5-oxotetrahydrofuran-2-yl)vinyl) isobenzofuran-1(3*H*)-one.

Table 2. ^{1}H NMR data (500 MHz) and ^{13}C NMR data (125 MHz) of 3 and 4.

No.	3 [a]		4 [b]	
	δ_C	δ_H	δ_C	δ_H
1	172.9 C		170.6 C	
3	70.0 CH$_2$	5.18, s	69.2 CH$_2$	5.30, s
3a	144.1 C		147.5 C	
4	116.7 C		117.0 C	
5	163.8 C		162.8 C	
6	122.5 C		117.9 C	
7	153.7 C		153.5 C	
7a	106.3 C		107.8 C	
8	11.6 CH$_3$	2.13, s	11.3 CH$_3$	2.09, s
9	61.1 CH$_3$	3.77, s	60.6 CH$_3$	3.67, s
1'	21.3 CH$_2$	2.66, m	118.2 CH	6.63, d (16.5)
2'	33.1 CH$_2$	1.90, m	137.4 CH	6.76, d (16.1)
		1.67, m		
3'	39.4 CH	2.50, dq (6.9, 6.9)	85.8 C	
4'	177.0 C		34.0 CH$_2$	2.24, m
				2.19, m
5'	51.6 CH$_3$	3.68, s	28.8 CH$_2$	2.64, m
				2.56, m
6'	17.1 CH$_3$	1.21 d (1.2)	177.0 C	
7'			26.4 CH$_3$	1.55, s

[a] CDCl$_3$ used as solvent. [b] DMSO-d_6 used as solvent.

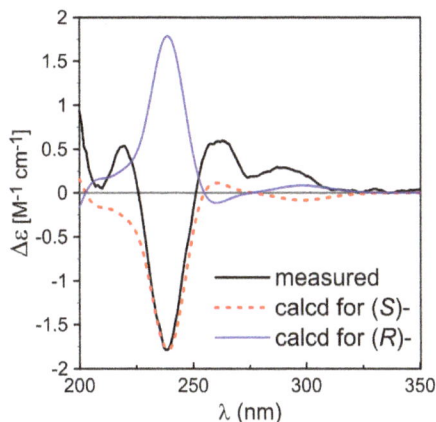

Figure 5. Comparison of the measured ECD spectrum of **4** with B3LYP/TZVP calculated spectra of (*S*)- and (*R*)-**4** in MeOH (σ = 0.20 eV, shift = −18 nm).

Compounds **5–10** were identified as 6-(3-carboxybutyl)-7-hydroxy-5-methoxy -4-methylphthala n-1-one [18], 7-hydroxy-6-[2-hydroxy-2-(2-methyl-5-oxotetrahydro-2-furyl) ethyl]-5-methoxy-4-meth yl-1-phthalanone [10], 5-hydroxy-7-methoxy-4 –methylphtalide [19], mycochromenic acid [10,20], (−)-brevianamide C [21], and (+)-brevianamide A [22], respectively.

Compounds **1–10** were tested for cytotoxicity against human colon cancer HCT116 cell line using sulforhodamine B (SRB) assay. Only Compound **9** exhibited moderate activity with an IC$_{50}$ value of 15.6 μM. In addition, antifouling activities of **1**, **3**, **5**, and **8** were also evaluated in settlement inhibition assays with *Bugula neritina* larvae. The larval settlement bioassay showed that **3** and **5** could strongly inhibit the larvae settlement of *B. neritina* larvae with EC$_{50}$ values of 13.7 and 22.6 μM, respectively, and $LC_{50}/EC_{50} > 100$, while **1** and **8** showed no antilarval activity. Moreover, **1–10** were evaluated for their anti-bacterial activities against *Streptococcus mutans* (ATCC35668) and *S. sobrinus* (ATCC33478), anti-fungal activities against *Fusarium oxysporum* f. sp. *cubense* Race 1 and Race 4; however, they did not show any obvious activity at the concentration of 20 μg/mL.

3. Experimental Section

3.1. General Experimental Procedure

Optical rotations were measured with an Anton Paar MCP 500 polarimeter (Anton Paar GmbH, Graz, Austria). UV spectra were measured with a Shimadzu UV-2600 UV–vis spectrophotometer (Shimadzu, Kyoto, Japan) in a MeOH solution. Infrared spectra (IR) were recorded on a Shimadzu IRAffinity-1 Fourier transform infrared spectrophotometer (Shimadzu, Kyoto, Japan). ^{1}H/^{13}C NMR and 2D NMR spectra were recorded on a Bruker AV-500 spectrometer (Bruker, Billerica, MA, USA) with TMS as reference. High-resolution electrospray-ionization (HRESIMS) was performed on a Bruker microTOF-QII mass spectrometer (Bruker, Bremen, Germany). Analysis HPLC was performed on an Angilent 1100 liquid chromatography system (Agilent Technologies, Santa Clara, CA, USA). Semi-preparative reversed-phase HPLC was performed on a Shimadzu LC-20A preparative liquid chromatography (Shimadzu, Kyoto, Japan) with YMC-Pack ODS-A column 250 × 10 mm i.d., S-5 μm × 12 nm, and 250 × 20 mm i.d., S-5 μm × 12 nm. Column chromatography (CC) was performed on silica gel (200–300 mesh, Qingdao Marine Chemical, Qingdao, China), Sephadex LH-20 (GE Healthcare, Barrington, IL, USA), or Rp-18 silica (Pharmacia Co. Ltd., St. Paul, MN, USA).

3.2. Fungal Material

The fungal strain DFFSCS025 (GenBank access number JX156371) was isolated from a deep-sea sediment sample collected in the South China Sea, Sansha City (18°5′ N, 118°31′ E; 3928 m depth), Hainan Province, and identified as *Penicillium brevicompactum* by ITS rDNA sequence homology (99% similarity with *P. brevicompactum*). The strain was deposited in the RNAM Center, South China Sea Institute of Oceanology, Chinese Academy of Sciences.

3.3. Fermentation and Extraction

Spores of the fungal strain were inoculated into 1000 mL Erlenmeyer flasks each containing 300 mL of liquid medium (glucose 1%, maltose 2%, monosodium glutamate 1%, KH_2PO_4 0.05%, $MgSO_4 \cdot 7H_2O$ 0.003%, corn steep liquor 0.05%, yeast extract 0.3%, dissolved in sea water, pH 6.5). After 32 days of stationary cultivation at 28 °C, the whole broths (30 L) were filtered through cheesecloth. Sterilized XAD-16 resin (20 g/L) was added to the liquor and shaken at low speed for 30 min to absorb the organic products. The resin was washed with distilled water to remove medium residue then eluted with methanol. The methanol solvent was removed under vacuum to yield a brown residue (18 g). The mycelium portion was smashed and extracted twice with 80% acetone/H_2O. The acetone soluble fraction was dried in vacuo to yield 6 g of residue. The residues of liquor and mycelium extracts were combined together according to TLC chromatography detecting.

3.4. Purification

The combined extract (24 g) was subjected to silica gel column (500 g) and eluted with $CHCl_3$/MeOH (100:0–80:20, v/v) to yield ten fractions (Fractions 1–10). Fraction 4 (10.5 g) was separated by silica gel column and eluted with $CHCl_3$/MeOH to give seven sub-fractions (Fractions 4-1–4-7). Fraction 4-4 (0.9 g) was subjected to Develosil ODS column eluting with a decreasing polarity of MeOH/H_2O (20:80–70:30) and purified with semi-preparation HPLC (MeOH/H_2O, 65:35) at the flow rate of 3 mL/min to yield **6** (t_R 68.2 min, 16.5 mg) and **7** (t_R 15.3, 2.4 mg). Fraction 6 was isolated with Develosil ODS column eluting with MeOH/H_2O (15:85–70:30) to obtain six sub-fractions (Fractions 6-1–6-6). Fraction 6-4 was purified by preparatory HPLC (CH_3CN/H_2O, 34:66) at the flow rate of 6 mL/min to yield **2** (t_R 18.4 min, 4.7 mg), **10** (t_R 25.5 min, 1.7 mg), **9** (t_R 28.4 min, 2.7 mg), **5** (t_R 35.8 min, 4.2 mg), and **4** (t_R 53.4 min, 6.0 mg), respectively. Fraction 6-5 was purified by HPLC (MeOH/H_2O, 65:35) at the flow rate of 6 mL/min to yield **1** (t_R 13.6 min, 3.2 mg), **8** (t_R 15.8 min, 55.3 mg), and **3** (t_R 27.1 min, 25.5 mg).

Brevianamide X (**1**): yellowish powder; $[\alpha]_D^{25}$ +8.7 (*c* 0.2, MeOH); UV (MeOH) λ_{max} (log ε) 209 (4.44), 253 (3.68), 286 (3.08) nm; CD (MeOH) λ_{max} ($\Delta\varepsilon$) 227 (-24.6), 252 (+10.1), 290 (+3.0); ^1H and ^{13}C NMR data, see Table 1; HRESIMS *m/z* 366.1810 [M + H]$^+$ (calcd. for 366.1812), 388.1630 [M + Na]$^+$ (calcd. for 388.1632).

Brevianamide Y (**2**): yellowish powder; $[\alpha]_D^{25}$ +11.5 (*c* 0.2, MeOH); UV (MeOH) λ_{max} (log ε) 209 (4.64), 252 (3.88), 286 (3.25) nm; CD (MeOH) λ_{max} ($\Delta\varepsilon$) 207 (-2.0), 218 (+4.5), 238 (-21.7), 262 (+7.1), 276 (+2.0), 293 (+3.1); ^1H and ^{13}C NMR data, see Table 1; HRESIMS *m/z* 366.1813 [M + H]$^+$ (calcd. for 366.1812), 388.1636 [M + Na]$^+$ (calcd. for 388.1632).

6-(Methyl 3-methylbutanoate)-7-hydroxy-5-methoxy-4-methylphthalan-1-one (**3**): white powder; $[\alpha]_D^{25}$ 0 (*c* 0.1, MeOH); UV (MeOH) λ_{max} (log ε) 209 (4.67), 286 (3.31), 313 (2.67) nm; CD (MeOH) λ_{max} ($\Delta\varepsilon$) ±0; ^1H and ^{13}C NMR data, see Table 2; HRESIMS *m/z* 309.1332 [M + H]$^+$ (calcd. for 309.1333), 331.1164 [M + Na]$^+$ (calcd. for 331.1152).

(3′S)-(E)-7-Hydroxy-5-methoxy-4-methyl-6-(2-(2-methyl-5-oxotetrahydrofuran-2-yl)vinyl)isobenzofuran-1 (3H)-one (**4**): white powder; $[\alpha]_D^{25}$ +2.2 (*c* 0.1, MeOH); UV (MeOH) λ_{max} (log ε) 209 (4.05), 244 (4.04), 334 (3.22) nm; CD (MeOH) λ_{max} ($\Delta\varepsilon$) 210 (+0.05), 220 (+0.54), 238 (-1.79), 261 (+0.59), 274 (+0.17), 290 (+0.29); ^1H ^{13}C NMR data, see Table 2; HRESIMS *m/z* 341.1005 [M + Na]$^+$ (calcd. for 341.0996).

7-Hydroxy-6-[2-hydroxy-2-(2-methyl-5-oxotetrahydro-2-furyl)ethyl]-5-methoxy-4-methyl-1-phthalanone (**6**): white powder; $[\alpha]_D^{25}$ -0.8 (*c* 0.1, MeOH); CD (MeOH) λ_{max} ($\Delta\varepsilon$) ±0; ^1H NMR (500 MHz, DMSO-d_6) δ_H: 5.24 (2H, s, H-3), 2.09 (3H, s, H-8), 3.75 (3H, s, H-9), 2.78 (1H, dd, *J* = 2.0, 13.5 Hz, H-1′α), 2.71 (1H, dd, *J* = 10.1, 13.5 Hz, H-1′β), 3.83 (1H, dd, *J* = 1.9, 9.8 Hz, H-2′), 2.37 (1H, ddd, *J* = 7.3, 10.0, 12.3 Hz, H-4′α), 1.87 (1H, ddd, *J* = 6.3, 10.3, 12.5 Hz, H-4′β), 2.61 (1H, dd, *J* = 7.3, 10.8 Hz, H-5′α), 2.56 (1H, dd, *J* = 6.4, 10.0 Hz, H-5′β), 1.41 (3H, s, H-7′); ^{13}C NMR (125 MHz, DMSO-d_6) δ_C: 170.23 (C, C-1), 68.85 (CH$_2$, C-3), 146.84 (C, C-3a), 116.26 (C, C-4), 163.50 (C, C-5), 120.99 (C, C-6), 154.31 (C, C-7), 107.48 (C, C-7a), 11.58 (CH$_3$, C-8), 61.13 (CH$_3$, C-9), 26.65 (CH$_2$, C-1′), 74.61 (CH, C-2′), 88.65 (C, C-3′), 28.58 (CH$_2$, C-4′), 29.35 (CH$_2$, C-5′), 177.13 (C, C-6′), 22.53 (CH$_3$, C-7′); HRESIMS *m/z* 335.1146 (M − H)$^-$ (calcd. for 335.1136).

Mycochromenic acid (**8**): white powder; $[\alpha]_D^{25}$ -3.1 (*c* 0.15, MeOH); ^1H NMR (500 MHz, DMSO-d_6) δ_H: 5.22 (2H, s, H-3), 2.07 (3H, s, H-8), 3.75 (3H, s, H-9), 6.62 (1H, d, *J* = 10.1 Hz, H-1′), 5.83 (1H, d, *J* = 10.1 Hz, H-2′), 1.93 (2H, m, H-4′), 2.35 (2H, t, *J* = 7.8 Hz, H-5′), 1.40 (3H, s, H-7′); ^{13}C NMR (125 MHz, DMSO-d_6) δ_C: 168.22 (C, C-1), 68.73 (CH$_2$, C-3), 148.95 (C, C-3a), 116.98 (C, C-4), 159.36 (C, C-5), 114.70 (C, C-6), 151.03 (C, C-7), 107.95 (C, C-7a), 11.02 (CH$_3$, C-8), 62.04 (CH$_3$, C-9), 117.52 (CH, C-1′), 130.02 (CH, C-2′), 79.19 (C, C-3′), 35.52 (CH$_2$, C-4′), 28.79 (CH$_2$, C-5′), 174.54 (C, C-6′), 26.10 (CH$_3$, C-7′); ESIMS *m/z* 363 (M + Na)$^+$.

(-)-Brevianamide C (**9**): yellowish powder; $[\alpha]_D^{25}$ −60.4 (*c* 0.2, MeOH); UV (MeOH) λ_{max} (log ε) 202 (4.02), 235 (4.02), 259 (4.13) nm; CD (MeOH) λ_{max} ($\Delta\varepsilon$) 205 (−16.40), 225 (+10.43), 244 (+1.22), 252 (+1.61), 265 (−2.15), 302 (+0.57); ^1H NMR (500 MHz, CDCl$_3$) δ_H: 7.70 (1H, d, *J* = 7.5 Hz, H-4), 6.93 (1H, t, *J* = 6.0 Hz, H-5), 7.45 (1H, td, *J* = 6.0, 6.0, 1.0 Hz, H-6), 6.91 (1H, d, *J* = 8.5 Hz, H-7), 5.84 (1H, s, H-10), 3.52 (2H, m, H-14), 2.04 (2H, dt, *J* = 6.5, 6.5, 12.7 Hz, H-15), 1.87 (1H, d, *J* = 15.4 Hz, H-16α), 2.80 (1H, dt, *J* = 6.9, 6.9, 13.4 Hz, H-16β), 1.83 (1H, dd, *J* = 6.0, 13.3 Hz, H-18α), 1.96 (1H, dd, *J* = 10.1, 13.3 Hz, H-18β), 2.46 (1H, m, H-19), 2.13 (1H, m, H-22), 0.85 (3H, d, *J* = 6.9 Hz, H-23), 0.86 (3H, d, *J* = 6.7 Hz, H-24); ^{13}C NMR (125 MHz, CDCl$_3$) δ_C: 186.76 (C, C-2), 139.03 (C, C-3), 125.60 (CH, C-4), 120.65 (CH, C-5), 136.80 (CH, C-6), 112.59 (CH, C-7), 154.55 (C, C-8), 121.62 (C, C-9), 104.03 (CH, C-10), 66.89 (C, C-11), 168.83 (C, C-12), 44.48 (CH$_2$, C-14), 24.55 (CH$_2$, C-15), 29.20 (CH$_2$, C-16), 66.21 (C, C-17), 30.83

(CH$_2$, C-18), 46.75 (CH, C-19), 172.07 (C, C-20), 27.67 (CH, C-22), 22.16 (CH$_3$, C-23), 16.17 (CH$_3$, C-24); ESIMS *m/z* 388 (M + Na)$^+$.

3.5. Computational Methods

Molecular Merck force field (MMFF) and DFT/TDDFT calculations were performed with Spartan'14 software package (Wavefunction Inc., Irvine, CA, USA) and Gaussian09 program package [23], respectively, using default grids and convergence criteria. MMFF conformational search generated low-energy conformers within a 10 kcal/mol energy window were subjected to geometry optimization using the B3LYP/def2-SVP method. Frequency calculations were run with the same method to verify that each optimized conformer was a true minimum and to estimate their relative thermal free energies (ΔG) at 298.15 K. Energies of the low-energy conformers in MeOH were calculated at the B3LYP/def2-TZVP level. Solvent effects were taken into account by using a polarizable continuum model (PCM). The TDDFT calculations were performed using the hybrid B3LYP [24–26] PBE1PBE [27,28] and/or TPSSh [29] functionals, and Ahlrichs' basis set TZVP (triple zeta valence plus polarization) [30]. The number of excited states per each molecule was 30. The ECD spectra were generated by the program SpecDis [31] using a Gaussian band shape from dipole-length dipolar and rotational strengths. Equilibrium population of each conformer at 298.15 K was calculated from its relative free energies using Boltzmann statistics. The calculated spectra were generated from the low-energy conformers according to the Boltzmann weighting of each conformer in a MeOH solution.

3.6. Cytotoxicity

All compounds were tested for cytotoxicity against HTC116 cell line with SRB method. Briefly, Cytotoxicity assays involving HCT116 cells were performed using sulforhodamine B based on slightly modified protocols [32]. HCT116 cells were maintained in a DMEM medium with 10% fetal bovine serum (FBS) (Life Technologies, Carlsbad, CA, USA). Tested samples were prepared using 10% aqueous DMSO as solvent. The cell suspension was added into 96-well microliter plates in 190 µL at plating densities of 5000 cells/well. One plate was fixed in situ with TCA to represent a no-growth control at the time of drug addition (Day 0). Then, 10 µL of 10% aqueous DMSO was used as control group. After 72 h incubation, the cells were fixed to plastic substratum by the addition of 50 µL of cold 50% aqueous trichloroacetic acid and washed with water after incubation at 4 °C for 30 min. After staining cells with 100 µL of 0.4% sulforhodamine B in 1% aqueous AcOH for 30 min, unbound dye was removed by washing four times with 1% aqueous AcOH. The plates were allowed to dry at room temperature, then the bound dye was solubilized with 200 µL of 10 mM unbuffered Tris base, pH 10. Shaken for 5 min or until the dye was completely solubilized and the optical density was measured at 515 nm using an ELISA plate reader (Bio-Rad, Hercules, CA, USA). The average data were expressed as a percentage, relative to the control. Percentage growth inhibition was calculated as (OD (cells + samples) − OD (Day 0 only cells))/(OD (cells + 10% DMSO) − OD (Day 0 only cells)) = % survival, cytotoxicity = 1 − % survival. (Graphpad Software, Inc., San Diego, CA, USA).

3.7. Larval Settlement Assays

Larval culture and larval settlement assays matched the method reported in the literature [33]. Briefly, the stock solution of tested samples in DMSO was diluted with autoclaved filtered sea water (FSW) to concentrations ranging from 3.125 to 100 µg/mL. About 20 competent larvae were added to each well in 1 mL of the test solution. Wells containing only FSW with DMSO served as the controls. The plates were incubated at 27 °C for 1 h for *B. neritina* and 24 h for *B. amphitrite*. The percentage of larval settlement was determined by counting the settled, live individuals under a dissecting microscope and expressing the result as a proportion of the total number of larvae in the well. EC$_{50}$ (inhibits 50% of larvae settlement in comparison with the control) and LC$_{50}$ (lethal concentration, 50%) were calculated by using the Excel software program.

4. Conclusions

In conclusion, four new compounds (**1–4**), include two brevianamides and two mycochromenic acid derivatives along with six known compounds were isolated from the deep-sea-derived fungus *Penicillium brevicompactum* DFFSCS025. Their structures were elucidated by spectroscopic analysis and quantum chemical computations. Compound **9** showed moderate cytotoxicity against human colon cancer HCT116 cell line with IC_{50} values of 15.6 μM. Compounds **3** and **5** had significant antifouling activity against *Bugula neritina* larval settlement with EC_{50} values of 13.7 and 22.6 μM, respectively. The NMR data of **6**, **8**, and **9** were assigned for the first time.

Supplementary Materials: Available online: www.mdpi.com/1660-3397/15/2/43/s1. ^{1}H, ^{13}C, and 2D NMR spectra and HRESIMS of Compounds **1–4** and known Compounds **6**, **8**, and **9**. HPLC spectra of Compounds **3** and **5**. Chiral HPLC spectra of Compounds **3** and **6**.

Acknowledgments: We thank X.-Y. Wei, South China Botanical Garden, Chinese Academy of Sciences, for quantum chemical computations of **1**, **2**, and **4**. We thank C.-F. Ku, School of Chinese Medicine, Hong Kong Baptist University, for anti-tumor activity determination. We are grateful for the financial support provided by the Regional Innovation Demonstration Project of Guangdong Province Marine Economic Development (GD2012-D01-002), the Natural Science Foundation of China (41606186), the Strategic Leading Special Science and Technology Program of Chinese Academy of Sciences (XDA100304002), the Natural Science Foundation of China (41376160 and 81673326), Guangzhou Science and Technology Research Projects (201607010305) and the National Marine Public Welfare Research Project of China (201305017).

Author Contributions: Shuhua Qi designed and guided the experiment. Xinya Xu purified and conducted the structure elucidation. Xiaoyong Zhang contributed to isolate the fungus from deep-sea sediment. Xuhua Nong had evaluated anti-fouling activity using *Bugula neritina* larvae. Jie Wang performed anti-bacterial activity and anti-fungal activity.

Conflicts of Interest: The authors declare no conflict of interest.

References

1. Skropeta, D. Deep-sea natural products. *Nat. Prod. Rep.* **2008**, *25*, 1131–1166. [CrossRef] [PubMed]
2. Skropeta, D.; Wei, L. Recent advances in deep-sea natural products. *Nat. Prod. Rep.* **2014**, *31*, 999–1025. [CrossRef] [PubMed]
3. Zhang, W.; Liu, Z.; Li, S.; Yang, T.; Zhang, Q.; Ma, L.; Tian, X.; Zhang, H.; Huang, C.; Zhang, S.; et al. Spiroindimicins A–D: New bisindole alkaloids from a deep-sea-derived Actinomycete. *Org. Lett.* **2012**, *14*, 3364–3367. [CrossRef] [PubMed]
4. Birch, A.J.; Wright, J.J. The brevianamides: A new class of fungal alkaloid. *J. Chem. Soc. D.* **1969**. [CrossRef]
5. Paterson, R.R.M.; Simmonds, M.J.S.; Kemmelmeier, C.; Blaney, W.M. Effects of brevianamide A, its photolysis product brevianamide D, and ochratoxin A from two *Penicillium* strains on the insect pests *Spodoptera frugiperda* and *Heliothis virescens. Mycol. Res.* **1990**, *94*, 538–542. [CrossRef]
6. Song, F.; Liu, X.; Guo, H.; Ren, B.; Chen, C.; Piggott, A.M.; Yu, K.; Gao, H.; Wang, Q.; Liu, M.; et al. Brevianamides with antitubercular potential from a marine-derived isolate of *Aspergillus versicolor. Org. Lett.* **2012**, *14*, 4770–4773. [CrossRef] [PubMed]
7. Qin, W.F.; Xiao, T.; Zhang, D.; Deng, L.F.; Wang, Y.; Qin, Y. Total synthesis of (-)-depyrannoversicolamide B. *Chem. Commun.* **2015**, *51*, 16143–16146. [CrossRef] [PubMed]
8. Williams, R.M.; Glinka, T.; Kwast, E.; Coffman, H.; Stille, J.K. Asymmetric, stereocontrolled total synthesis of (-)-brevianamide B. *J. Am. Chem. Soc.* **1990**, *112*, 808–821. [CrossRef]
9. Zhao, L.; May, J.P.; Huang, J.; Perrin, D.M. Stereoselective synthesis of brevianamide E. *Org. Lett.* **2012**, *14*, 90–93. [CrossRef] [PubMed]
10. Jones, D.F.; Moore, R.H.; Crawley, G.C. Microbial modification of mycophenolic acid. *J. Chem. Soc. C.* **1970**. [CrossRef]
11. Sintchak, M.D.; Fleming, M.A.; Futer, O.; Raybuck, S.A.; Chambers, S.P.; Caron, P.R.; Murcko, M.A.; Wilson, K.P. Structure and mechanism of inosine monophosphate dehydrogenase in complex with the immunosuppressant mycophenolic acid. *Cell* **1996**, *85*, 921–930. [CrossRef]
12. Bentley, R. Mycophenolic acid: A one hundred year odyssey from antibiotic to immunosuppressant. *Chem. Rev.* **2000**, *100*, 3801–3825. [CrossRef] [PubMed]

13. Birch, A.J.; Wright, J.J. Biosynthesis. XLII. Structural elucidation and same aspects of the biosynthesis of the brevianamides-A and -E. *Tetrahedron* **1970**, *26*, 2329–2344. [CrossRef]

14. Nugroho, A.E.; Morita, H. Circular dichroism calculation for natural products. *J. Nat. Prod.* **2014**, *68*, 1–10. [CrossRef] [PubMed]

15. Miao, F.P.; Liang, X.R.; Liu, X.H.; Ji, N.Y. Aspewentins A-C, norditerpenes from a cryptic pathway in an algicolous strain of Aspergillus wentii. *J. Nat. Prod.* **2014**, *77*, 429–432. [CrossRef] [PubMed]

16. Castro, V.; Jakupovic, J.; Bohlmann, F. A new type of sesquiterpene and acorane derivative from *Calea prunifolia*. *J. Nat. Prod.* **1984**, *47*, 802–808. [CrossRef]

17. Mohapatra, D.K.; Pramanik, C.; Chorghade, M.S.; Gurjar, M.K. A short and efficient synthetic strategy for the total syntheses of (*S*)-(+)- and (*R*)-(−)-plalolide A. *Eur. J. Org. Chem.* **2007**, 5059–5063. [CrossRef]

18. Habib, E.; Leon, F.; Bauer, J.D.; Hill, R.A.; Carvalho, P.; Cutler, H.G.; Cutler, S.J. Mycophenolic derivatives from *Eupenicillium parvum*. *J. Nat. Prod.* **2008**, *71*, 1915–1918. [CrossRef] [PubMed]

19. Valente, A.M.M.P.; Ferreira, A.G.; Daolio, C.; Filho, E.R.; Boffo, E.; Souza, A.Q.L.; Sebastianes, F.L.S.; Melo, I.S. Production of 5-hydroxy-7-methoxy-4-methylphthalide in a culture of *Penicillium crustosum*. *An. Acad. Bras. Cienc.* **2013**, *85*, 487–496. [CrossRef] [PubMed]

20. Yamaguchi, S.; Nedachi, M.; Maekawa, M.; Murayama, Y.; Miyazawa, M.; Hirai, Y. Synthetic study for two 2*H*-chromenic acids, 8-chlorocannabiorcichromenic acid and mycochromenic acid. *J. Heterocycl. Chem.* **2006**, *43*, 29–41. [CrossRef]

21. Birch, A.J.; Russell, R.A. Structural elucidations of brevianamides-B, -C, -D and -F. *Tetrahedron* **1972**, *28*, 2999–3008. [CrossRef]

22. Williams, R.M.; Sanz-Cervera, J.F.; Sancenon, F.; Marco, J.A.; Halligan, K.M. Biomimetic Diels-Alder cyclizations for the construction of the brevianamide, paraherquamide, sclerotamide, asperparaline and VM55599 ring systems. *Bioorg. Med. Chem.* **1998**, *6*, 1233–1241. [CrossRef]

23. Frisch, M.J.; Trucks, G.W.; Schlegel, H.B.; Scuseria, G.E.; Robb, M.A.; Cheeseman, J.R.; Scalmani, G.; Barone, V.; Mennucci, B.; Petersson, G.A.; et al. *Gaussian 09*; Revision C.01; Gaussian, Inc.: Wallingford, CT, USA, 2010.

24. Becke, A.D. Density-functional thermochemistry. III. The role of exact exchange. *J. Chem. Phys.* **1993**, *98*, 5648–5652. [CrossRef]

25. Becke, A.D. Density-functional exchange-energy approximation with correct asymptotic-behavior. *Phys. Rev. A Gen. Phys.* **1988**, *38*, 3098–3100. [CrossRef] [PubMed]

26. Lee, T.; Yang, W.T.; Parr, R.G. Development of the Colle-Salvetti correlation-energy formula into a functional of the electron density. *Phys. Rev. B Condens. Matter* **1988**, *37*, 785–789. [CrossRef] [PubMed]

27. Perdew, J.P.; Burke, K.; Ernzerhof, M. Generalized gradient approximation made simple. *Phys. Rev. Lett.* **1996**, *77*, 3865–3868. [CrossRef] [PubMed]

28. Adamo, C.; Barone, V. Toward reliable density functional methods without adjustable parameters: The PBE0 mode. *J. Chem. Phys.* **1999**, *110*, 6158–6169. [CrossRef]

29. Staroverov, V.N.; Scuseria, G.E.; Tao, J.; Perdew, J.P. Efficient hybrid density functional calculations in solids: Assessment of the Heyd-Scuseria-Ernzerhof screened Coulomb hybrid functional. *J. Chem. Phys.* **2003**, *119*, 12129. [CrossRef]

30. Schäfer, A.; Huber, C.; Ahlrichs, R. Fully optimized contracted gaussian basis sets of triple zeta valence quality for atoms Li to Kr. *J. Chem. Phys.* **1994**, *100*, 5829–5835. [CrossRef]

31. Bruhn, T.; Schaumlöffel, A.; Hemberger, Y.; Bringmann, G. SpecDis: quantifying the comparison of calculated and experimental electronic circular dichroism spectra. *Chirality* **2013**, *25*, 243–249. [CrossRef] [PubMed]

32. Li, W.F.; Wang, J.; Zhang, J.J.; Song, X.; Ku, C.F.; Zou, J.; Li, J.X.; Rong, L.J.; Pan, L.T.; Zhang, H.J. Henrin A: A new anti-HIV ent-kaurane diterpene from *Pteris henryi*. *Int. J. Mol. Sci.* **2015**, *16*, 27978–27987. [CrossRef] [PubMed]

33. Qi, S.H.; Xu, Y.; Xiong, H.R.; Qian, P.Y.; Zhang, S. Antifouling and antibacterial compounds from a marine fungus *Cladosporium* sp. F14. *World J. Microbiol. Biotechnol.* **2009**, *25*, 399–406. [CrossRef]

MDPI AG

St. Alban-Anlage 66

4052 Basel, Switzerland

Tel. +41 61 683 77 34

Fax +41 61 302 89 18

http://www.mdpi.com

Marine Drugs Editorial Office

E-mail: marinedrugs @mdpi.com

http://www.mdpi.com/journal/marinedrugs

www.ingramcontent.com/pod-product-compliance
Lightning Source LLC
Chambersburg PA
CBHW051914210326
41597CB00033B/6133

* 9 7 8 3 0 3 8 4 2 7 6 5 0 *